Security, Rights, and Liabilities in E-Commerce

For a complete listing of the *Artech House Telecommunications Library,*
turn to the back of this book.

Security, Rights, and Liabilities in E-Commerce

Jeffrey H. Matsuura

Artech House
Boston • London
www.artechhouse.com

Library of Congress Cataloging-in-Publication Data
Matsuura, Jeffrey H., 1957–
    Security, rights, and liabilities in E-commerce / Jeffrey Matsuura.
        p.   cm.—(Artech House telecommunications library)
        Includes bibliographical references and index.
        ISBN 1-58053-298-5 (alk. paper)
        1. Electronic commerce—Law and legislation—United States.   2. Security
    (Law)—United States.   3. Data protection—Law and legislation—United States.
    4. Database security—United States.   I. Title.   II. Series.
        KF889 .M359 2001
        343.7309'944—dc21   2001045737

British Library Cataloguing in Publication Data
Matsuura, Jeffrey
    Security, rights, and liabilities in E-commerce. — (Artech House
    telecommunications library)
    1. Electronic commerce—Law and legislation   2. Data protection—Law and
    legislation
    I. Title
    343' .0999
    ISBN 1-58053-298-5

Cover by Yekaterina Ratner

© 2002 ARTECH HOUSE, INC.
685 Canton Street
Norwood, MA 02062

International Standard Book Number: 1-58053-298-5
Library of Congress Catalog Card Number: 2001045737

10 9 8 7 6 5 4 3 2 1

This book is dedicated to Harry and Tomeko Matsuura, and to Janice, Anne, and Fuzz, in appreciation for the lessons they have taught me in the past, and those that they will share with me in the days to come.

Contents

1

Overview: Legal Aspects of Security in the Digital Marketplace

This book examines some of the fundamental legal rights and liabilities associated with electronic commerce security. It places those rights and liabilities into the context of the digital marketplace, describing some of the more important legal issues associated with security provided for the most common on-line commercial transactions and relationships. The book is intended to help readers understand legal rights that they may be able to enforce and legal obligations that they will be required to satisfy with respect to security, as they conduct business in the digital marketplace.

As you read this book and consider the issues it raises, always remember that legal rights and liabilities are two different sides of the same coin. One party's legal rights commonly translate into another party's legal obligations. Each time another party fails to meet its legal duties, your rights may be infringed upon, and conversely each time you fail to meet one of your responsibilities, that failure may infringe on the rights of another. Legal compliance is an effort to avoid liabilities, but it should also be viewed as an investment in conduct that we hope will encourage others to act responsibly, thereby protecting our own rights in the long run.

What is the digital marketplace?

The digital marketplace is the global network of commercial transactions and economic relationships supported by the Internet and other forms of modern information and telecommunications technology. Described another way, the digital marketplace is the full range of electronic commerce functions and processes, encompassing both business-to-consumer and business-to-business activities. The digital marketplace also includes all of the electronic transactions and information processing performed to support and execute traditional (e.g., "brick and mortar") commercial activities. In the digital marketplace, goods in electronic form are created and distributed. The digital marketplace also provides an important part of the infrastructure to support creation and distribution of tangible goods. For our purposes, the digital marketplace is thus the combination of the world of "clickstream commerce" and the "bricks and clicks" commercial environment.

The digital marketplace is becoming a major component of the overall economy. When we combine purely electronic transactions with traditional commercial transactions that are supported by the Internet and other computer systems, as well as financial transfers (e.g., bank-to-bank funds transfers) that are performed electronically, we see that the digital marketplace already comprises a significant percentage of the total global economic activity. There is every reason to believe that the size of the digital marketplace will continue to increase dramatically for the foreseeable future.

What are the sources of the legal rights addressed in this book?

The legal rights and obligations discussed in this book arise from a number of sources. Some of those rights are created by laws or statutes that are enacted by legislative bodies. Other rights are created by the courts ("common law") as they make decisions resolving specific disputes. Certain obligations are created at the international level through treaties. Additional rights are created by the development of rules, enacted and enforced by regulatory or administrative agencies (e.g., the Federal Communications Commission in the United States). Some standards are established by private groups, such as industry associations. Although these standards do not have the force of law, they can have a significant impact on conduct to the extent that the private groups that create and enforce them control or influence the behavior of participants in the marketplace.

Several different substantive categories of law apply to digital security. Intellectual property law (e.g., copyright and trademark), trade secrets law, and privacy law, for example, have an important impact on rights and duties associated with e-commerce security. The law of contracts and commercial transactions also plays a key role in digital security. Increasingly, consumer protection law, the law of torts, antitrust law, and property law are influencing digital marketplace security. Finally, we must always recognize that criminal law has an important role to play in digital security.

The rights and obligations discussed in this book are generally enforced by two different groups. Some of these legal rights are enforced by private parties. Businesses or individual people have the right to enforce certain legal obligations against other parties (e.g., contract law rights). Legal actions to enforce these rights are sometimes referred to as civil actions or private law claims. Other legal rights are enforced by governments. Local, regional or national governments have the authority to enforce criminal laws and administrative rules (e.g., regulatory laws). Certain legal rules can be enforced by both governments and private parties (e.g., antitrust laws).

What is the purpose of this book?

The primary purpose of this book is to help practitioners involved in the digital marketplace understand some of the more pressing legal issues associated with electronic commerce security. With that understanding of the issues, it is hoped that those practitioners will be better able to avoid legal liability while protecting their own rights in the future. That understanding of the issues should also help readers make more efficient use of their attorneys and other providers of professional services whom they consult. An awareness of some of the most significant legal aspects of digital commerce security should help the reader to anticipate problems and to take steps to avoid them. Awareness should also help readers deal more effectively with the security problems they actually encounter, enabling those readers to resolve their immediate security problems more quickly and with less risk. The book is designed to provide accurate and useful general information to its readers; however, it is not intended to offer direct legal or other professional services. Advice on specific legal issues should be sought from a professional who is familiar with the details of the specific context in which those issues have arisen.

To be most useful as a guide for digital marketplace participants, this book provides an overview of relevant issues and a summary of possible actions to address the issues. It does not offer an exhaustive or comprehensive analysis of those issues or actions. Curious readers seeking greater understanding of the issues highlighted in this text are, of course, welcome to consult the resources identified in the bibliography, which include relevant laws and court cases. Such additional research by the reader can provide helpful information, but should always be coupled with prudent consultation with attorneys and other professional service providers before being applied to any specific factual situation.

Who should read this book?

This book can provide useful information to anyone who is involved in electronic commerce and to anyone who is interested in the digital marketplace. Developers, owners, operators, and users of e-commerce systems and software form one set of readers. This group includes all of the parties involved in the creation, operation, and use of commercial Web sites, business-to-business commercial trading exchanges, on-line auctions systems, Internet service provider systems, and commercial computer networks. The group also includes providers of outsourced services supporting electronic commerce activities, such as Web-hosting service providers, application service providers, privacy and security providers, and transaction payment processors.

This book can also provide useful information to people involved in the creation and enforcement of public policy and to the general public. As the digital marketplace becomes an increasingly integral part of the economy, more consumers have a stake in the structure and operations of that marketplace. In that environment, consumers themselves have an interest in understanding more fully their legal rights and responsibilities, as they relate to electronic commerce security. Note that the category of consumers includes both individual people and businesses, all of whom are consumers in the digital marketplace. As the impact of the digital marketplace expands, lawmakers, regulators, and the courts must enhance their understanding of the rights and duties associated with security in that marketplace. This book can help the interested public and public policy professionals understand more clearly some of the key current legal issues associated with e-commerce security.

How is the book organized?

The book examines several important categories of legal issues associated with security for electronic commercial relationships and transactions. The legal significance of electronic documents and records is examined, particularly from the perspective of their importance as legally recognized evidence and documentation of conduct. This book also examines rights and duties associated with unauthorized access to computer systems and their content. The issue of the legal aspects of access management is addressed from the perspective of the need to prevent access by unauthorized system users and from the perspective of the need to control system access by authorized users to ensure that they limit their use of the system to the scope of the authority they were granted.

The book also reviews some of the key legal concerns associated with management of computer system content. The discussion of this issue focuses on the rights and duties associated with effective management of intellectual property and on the developing requirements for the protection of privacy for personal information. Also addressed are some of the key legal aspects of e-commerce transaction integrity, including important issues related to digital signatures, electronic contracts, and certification authorities.

The book also examines legal issues associated with security in the context of three increasingly popular digital marketplace practices: Internet auction transactions, on-line commercial trading exchanges (business-to-business commercial portals or marketplaces), and outsourcing of e-commerce functions. Use of all three of these practices is expanding rapidly, making those practices important aspects of the digital marketplace. Effective management of the security of the content and transactions associated with those systems is essential, as there are several important areas of potential legal liability that have been exposed by the broader use of auctions, exchanges, and outsourced services.

This book concludes with an overview of some of the future trends that are most likely to influence security in the digital marketplace tomorrow. It summarizes some of the most important legal implications of those trends, and also attempts to provide readers with basic principles and guidelines to help them prepare to deal with those trends and their security implications.

Basic lessons

Readers usually look for the basic lessons or principles that they should carry with them from the material that they review. When we read anything, we are generally looking for the fundamental themes that we will take away from the document and will retain for future application. For this book there are three core principles for the reader to consider.

The first principle is that security in the digital marketplace must be one of the primary objectives of all participants in that marketplace. Failure to devote that level of attention to e-commerce security will thwart the growth of the digital marketplace. There will be fewer users and fewer applications in the digital marketplace if its security is suspect. In addition, failure to focus on security for electronic commerce will lead to greater legal liability. If security is not a high priority, security breaches will be common, and those breaches will result in liability for many participants. Without effective security, there will be no true digital marketplace, and one of the basic requirements necessary to establish the requisite level of security is a recognition that e-commerce security is really a business management issue. Basic resources of the business (e.g., technology, people, capital) must be properly managed to provide security. It is a mistake to treat e-commerce security strategy as a set of decisions distinct from the overall business strategy of an organization. Digital security should now be a fundamental element of each enterprise's overall business strategy and planning.

The second principle is that the challenge of legal compliance will be substantially more complicated in the future than it is today. As more business transactions and relationships move to the electronic commerce environment and more parties participate in the digital marketplace, there will inevitably be more commercial conflicts, many of which will evolve into legal disputes. The increasing volume of users and transactions will also lead to greater regulation by governments, in virtually all jurisdictions.

This setting of more uses, more users, and more regulation for the digital marketplace will unquestionably make legal compliance for all participants in that marketplace a far greater challenge than it is today. All who would participate in the digital marketplace should thus be warned that the cost of participation will likely be greater tomorrow than it is today, as the costs associated with legal disputes and the costs of legal compliance will increase. It is possible that this increased cost of conducting business in the digital marketplace will work to the competitive advantage of large, well-established businesses that already have the resources and the experience required for complex, multijurisdictional legal-compliance activities.

The final principle is that legal compliance should be viewed as both a duty and an investment toward a future reward. Failure to meet your security obligations can result in penalties or other forms of legal liability. There is thus a set of legally enforceable obligations to act responsibly with regard to e-commerce security. But as noted previously, there is a symmetry between duties and rights. The law enforces duties to protect the rights of others. We are all simultaneously potential enforcers and infringers of legal rights. When we comply with legal obligations applied to us, we protect the rights of other parties.

All participants in the digital marketplace have an interest in making that marketplace a secure one. By complying with our legal duties to promote security, we are directly protecting the legal rights of other parties. We thus have a choice regarding how we think about our security obligations. We can, of course, consider them to be purely duties that we are responsible for undertaking, and that are subject to legal penalties if we fail to do so. This is an accurate assessment of those obligations. We might also, however, choose to think of our security actions as an investment in overall security for the marketplace. As we contribute to the security of the digital marketplace, perhaps we indirectly protect our own rights by providing others with greater incentive to fulfill their security duties, actions that help to protect our own rights. Viewed in this way, one could characterize legal compliance with security obligations as an investment that may help to yield, as a return, greater future protection for our security interests. By meeting our legal duties, we protect the rights of others, and by protecting the rights of others, we create an environment that provides greater protection for our own rights.

2

Managing Electronic Records and Evidence

Every electronic file, document, and transaction history is a record, a piece of evidence that can be used to verify information or to demonstrate that a specific transaction, in fact, took place. This can be good news, if the record helps us to prove a factual point we want to make; or it can be bad news, if it enables another party to prove a point against our interests. Most of us are familiar with the need to manage our printed documents, such as letters, memoranda, and reports to ensure that we know what material is in our records and what inferences can be drawn from those records. Many people are not as familiar with the need to manage their electronic records.

Failure to understand and manage electronic records can, however, have serious adverse consequences. Keep in mind that the law assumes that we are aware of all the information contained in our records, even if that is not actually the case. It makes us responsible for the contents of our records, and also requires us to disclose the contents of our records for public review, when ordered to do so by a court or other law enforcement authority. Failure to comply with these disclosure requirements can result in penalties imposed by governments. Those penalties include court-ordered compliance, fines, and prison terms, in some cases.

Certain types of electronic records are subject to special legal protection. For example various laws now require special protection for health and medical records in electronic form (e.g., the Health Insurance Portability and Accountability Act [HIPAA] in the United States). These specific categories of records are provided special legal protection as they contain material that is deemed by governments to be particularly sensitive. Parties who handle those special types of records must comply with the specific security requirements associated with those records, or face legal liability enforced by governments and by individuals who are the subjects of the records in question.

Failure to manage electronic records effectively can lead to legal liability to other private parties. Electronic records can be used by other parties to substantiate claims for court-ordered relief, such as monetary damages or court injunctions. Poorly managed electronic records may also undermine a party's ability to assert its own legal rights. A party may have been the target of actions that would ordinarily justify some form of compensation or other type of legal relief. If, however, that party is unable to provide evidence sufficient to support its claim for relief, it will be unable to assert its legal rights effectively. The party's electronic records will likely be an important element of its efforts to enforce its rights.

Organizations should develop strategies, policies, and practices that enable them to manage their general electronic records in a manner that helps them protect the value of those records as legal evidence. We should recognize that those records are subject to special legal requirements and ensure that the management systems applied to those special records comply with the legal obligations associated with them. Many forms of electronic records are relatively easy to identify and to manage (e.g., word processing documents, electronic mail messages). As our computer systems become more sophisticated, however, the range and complexity of electronic records increase dramatically. The list of the electronic records of many organizations now includes: cookies, cached on-line content, data tracking Web site usage, encryption keys, key-stroke monitoring data, and electronic commerce transaction audit data.

The challenge of performing all necessary records management functions effectively grows as the number of electronic records increases and as additional categories of electronic information are subject to special legal oversight. Technology and the demands of business competition provide the opportunity and the incentive to retain more and more information in electronic format. Each time we have an opportunity to retain additional electronic records, we should first consider the potential legal impact of possession of those records and make a conscious decision as to whether the commercial benefits of that additional material outweigh the potential liabilities associated with the material.

Records as evidence

When there is a legal dispute, there is a search for evidence. Evidence consists of information that helps a court or other legal institution (e.g., a law enforcement authority or a regulatory agency) to identify facts relevant to the dispute. Parties to the dispute produce evidence that helps clarify the facts, and based on that evidence, the judge or other party responsible for resolving the dispute makes a decision. Records—including electronic records—are a key component of evidence. Another important form of evidence is testimony, which consists of verbal or written statements and responses to specific questions, made under oath (i.e., subject to legal penalties for perjury if they are untrue), that provide information to the legal authority involved in the case.

Evidence is gathered for a legal proceeding through a process called *discovery*. The discovery process is the system through which relevant evidence is disclosed to the parties and the legal institutions involved in the case. Records, documents, and testimony are all gathered during discovery. The information collected during the discovery phase of the litigation creates the factual basis for the resolution of the dispute.

Courts, legislatures, and regulatory agencies have the authority to order private parties to make evidence available. These orders are commonly referred to as *subpoenas*. If you receive a court order to deliver information, including records, you must provide the information to the party requesting it at the time identified for delivery. If you fail to comply with a valid court order, you may be fined or jailed for *contempt* of the legal authority in question. The only way to avoid disclosing the information required by the court order is to persuade the authority that all or part of the sought material qualifies for one of the exemptions against disclosure established by law. Those exemptions include reasons such as *privilege* (e.g., the information requested is confidential communication between an attorney and his or her client, and is thus protected by *attorney-client privilege*). Claims of privilege against disclosure must, however, be approved by the authority that issued the order before they will be valid.

Legal systems in the United States and other countries give courts and other legal institutions the power to compel disclosure of evidence because that information is essential to the ability of those authorities to find the facts necessary to support fair rulings. Left to our own preferences, each of us would likely disclose information selectively. We would be eager to disclose information that supported our arguments, but would be reluctant to divulge evidence that weakened our case. In that setting, authorities would have difficulty making fair decisions; thus we give those authorities the potent and important tool of compelling disclosure of evidence.

When we receive a valid legal order to produce evidence, we cannot ignore it. More accurately, if we choose to ignore the order we will be subject to legal penalties. We must respond in some way, if we want to avoid legal liability for failure to comply. This obligation to produce records and evidence in response to a valid legal order exists in all jurisdictions, and it is an essential element of an effective legal system. The acceptable responses are: full disclosure of the material requested, a request for clarification of the information sought (or for additional time to comply), or a request that the court modify the order based on a claim that all or part of the sought material is, by law, exempt from disclosure. Notice that any action other than full compliance with the order must be approved by the legal authority. If the authority does not approve our request, we must comply fully with the original order, and if we do not so comply, we face penalties for contempt.

In legal actions in which the government is a party, the subpoenas are issued directly by the government. For example, in a criminal law case, the government issues the orders to compel disclosure of information. In civil lawsuits (cases in which one private party is suing another private party) the process is a bit different. Each party in a civil lawsuit makes its own request to the other party for disclosure of information. The party receiving a request is expected to comply with that request unless it can persuade the court (or regulatory agency) that is handling the case that such compliance is not required by law. In a civil case, the party that received the request must either comply or go to the court and ask to be excused from compliance. Compliance is required unless the court grants an exemption.

Recognize the difference between a compulsory order to disclose records and other information (e.g., a subpoena issued by a court, a legislature, or a regulatory agency) and a request for information (e.g., an inquiry made by a law enforcement authority or a private party). Compliance with valid orders issued by courts or other government bodies that have subpoena power is compulsory. Compliance with disclosure requests made by parties without subpoena power, including law enforcement authorities in many jurisdictions, is voluntary. You must comply with a compulsory legal order, and you may, if you choose to do so, comply with a request for information. Of course, a request for information can become a compulsory order if the requesting party can persuade a court to make the request mandatory. It is generally a good idea to decide ahead of time which requests for information, if any, your organization will accommodate. A common and sound approach to this type of disclosure is to adopt a policy that says the organization will only disclose records when required to do so by a compulsory court order, such as a subpoena.

Subpoenas for records are often broad. The authority commonly asks for "all documents and other records" related to a particular event or activity. When an order asks for "all" records, it means all records, no matter what form those records may take or how difficult they may be to produce. For example, in a breach of contract lawsuit, a subpoena may ask for all records of transactions between the parties during a specific period of time. To comply with that subpoena, the parties would be required to deliver all paper and electronic records documenting those transactions. Those records would include copies of paper forms, but they would also include all electronic data files documenting the electronic transactions between the parties during the relevant period of time. It would be possible to ask that the request be made more specific, that additional time be provided to comply, or that some of the material requested be exempt from disclosure, based on law, but it would not be reasonable to ignore the order.

Governments and private parties around the world recognize the significance of electronic records. Law enforcement authorities routinely seek court orders to seize and review computers and electronic files as part of many of their investigations. This practice is now so common that it is being integrated into new laws and regulations applicable to computer systems and their content. For example, the draft Convention on Cyber-Crime, a proposed treaty now being considered by the Council of Europe, provides for liberal access to the content of computers and computer networks by law enforcement authorities in European countries, making it easier for those authorities to search and seize computer equipment and stored content.

Private parties engaged in civil litigation also request electronic records as part of their standard discovery and document production processes. You can be certain that if you become involved in some form of legal action, the electronic records of your organization will be among the first targets of your adversary. For example, imagine that your organization has been sued by another party claiming that your organization engaged in unfair business practices. To be successful, the plaintiff in that case would like to have access to all of the internal communications (e.g., e-mail messages, memoranda) of your organization, to look for evidence regarding your group's actions, intentions, and motivations. It is likely that the plaintiff will make a very broad discovery request.

If your organization is like many other businesses, the records in question will likely contain vast amounts of potential evidence, some of which will help your case, some of which will hurt it, and much of which will not be familiar to you and your colleagues. That situation is not a good one from the perspective of protecting your legal interests, yet it is a very common situation. To make matters worse, remember that the information that a party obtains through the evidence discovery process can generally be retained and used by the party who discovers it for purposes other than the litigation. So in our example, the records obtained by the plaintiff through the discovery process could, unless the court orders special protection for them, be used by the plaintiff for other purposes (e.g., learning more about the way your organization operates or deriving information about your current commercial condition).

Obviously that information would be quite sensitive if the party who discovers it is a business competitor, supplier, or customer of your company. This situation explains, at least in part, why businesses and individuals sometimes initiate litigation, even if they are not confident that they will ultimately prevail. Under some circumstances the potential access to valuable competitive information, through the discovery process, may be attractive enough for commercial reasons to lead a party to go to court and bear the costs of litigation even with a relatively weak case. Courts are, of course, aware of this potential problem, and they commonly try to limit these so-called "fishing expeditions," but the risk remains a significant one. Effective management of records, in anticipation of litigation, is thus an essential process from both legal and business strategy perspectives. This challenge will increase significantly as the number of different types of electronic records increases, and parties involved in legal disputes become ever more eager to access electronic records to search for information that is valuable as legal evidence and commercial intelligence.

Consequences of disclosure

After records have been disclosed in response to a valid legal order, they become part of the evidence to be used to resolve the dispute. The party who supplied the records is held accountable for their accuracy. If those records have been altered or are false, the party who provided them to the court could be subject to legal penalties (i.e., fines, imprisonment) for falsifying evidence or misleading the court.

If the records sought by a legal authority contain information about third parties, additional caution is merited. Several court cases have, for example, raised fact patterns similar to the following. An Internet service provider (ISP) obtains information about one of its customers, and includes that information in its records. A legal action is brought against the customer by another party (e.g., a private party sues that customer for defamation), and as part of that litigation, a subpoena is issued to the ISP requesting all of the information the ISP has regarding that customer. The ISP provides the information in compliance with the subpoena. The customer then sues the ISP for breach of contract, claiming that the disclosure of the personal information violated the terms of service the ISP offered to the customer.

As a general rule, disclosure of information about third parties made in response to a valid order from a court or other legal institution is permissible and would not create liability to the disclosing party. The applicable theory is that the disclosure was required by law and is thus not the basis for a valid claim by the party who was the subject of the information. Be careful, however. In this type of situation, it is essential that the information not be disclosed before a valid legal order has been received. In instances when, for example, a company discloses such information in response to a mere request from a private party or a government agency, that disclosure could be the basis for liability. Make sure that a formal and valid legal order has been issued before making that type of disclosure.

Also be sure that your contractual commitments with customers and business partners reflect the fact that records will be disclosed in response to valid legal orders. All contracts containing confidentiality or nondisclosure provisions should specifically exempt material that is the target of a valid legal order compelling disclosure. Terms of service between your organization and its customers should also clearly indicate that information about the customer will be disclosed in response to valid legal orders. You should make sure that your business commitments do not promise more than you can deliver with regard to confidentiality for records that are the subject of court-ordered disclosure.

If the information sought pertains to a party with whom you have a special relationship (e.g., a customer, a business partner), additional steps may be appropriate prior to disclosure. You may want to challenge the order, requesting that all or part of the information be exempt from disclosure. This approach should only be taken if you have a good faith belief that there is a legal basis to exempt the information from disclosure or if you believe the request for information is overly broad. You may also want to inform the other party of the fact that the order has been issued, giving that party notice of the order's existence and of your anticipated response to it.

The example above involved an ISP, but other service providers now face similar issues of mandatory disclosure of customer records. For example, Web site–hosting service providers and outsourcing service providers handling e-mail, applications, and databases now commonly create and maintain electronic records for their clients. Those records are potentially subject to mandatory disclosure in a legal proceeding. Many different parties have an interest in such disclosure (i.e., the service provider, the client who owns the records, the party whose activities are reflected in the records, the parties to the legal proceeding, and the court), and those interests are often conflicting.

Imagine, for instance, a case involving a contract dispute between a company and the outsourced service provider it retained to manage the processing of electronic commerce transactions through the company's Web site. In that case, the company (the client) and the service provider would have conflicting legal interests. It is likely, however, that at least some of the electronic records important to the resolution of the dispute would be in the possession and under the control of the service provider, although they would presumably be the property of the client. This type of potentially conflicting interests between the party who owns the records and the party who possesses and controls them can lead to problems during litigation. For example, if the records help support the claims of the service provider, you can be sure that the service provider will move quickly and efficiently to provide them to the court. If those records bolster the case of the client, however, you can be equally certain that they will take much longer to be produced, and depending on the ethics and the attitudes toward risk of the service provider, those documents may end up being "lost" or "forgotten."

Law enforcement authorities and attorneys are acutely aware of the fact that electronic records are often located in many different places and are commonly not controlled by the parties who actually own them. As a result, subpoenas and requests for records are generally sent to any party who may have access to copies of those materials. If a party who has interests that conflict with the target party controls some of those records, it is good news for those who want access to the records, as the party possessing them is likely to surrender them without a fight. The path of least resistance when you are looking for access to records is to find someone who has a copy of those records and has no interest in keeping them confidential. This is a potential risk that we must all be aware of when we permit other parties to create, manage, and possess electronic records for us, but it is also a potential opportunity if we need access to the records of another party.

If you are a provider of information technology outsourcing services that include some form of electronic records development or management for your clients, you are in a particularly tricky situation regarding disclosure of those records. If faced with a valid legal order, you will be required to disclose those records. Often, this disclosure will upset your clients. One way to attempt to manage the situation is through the contract terms you enter into with your clients. Those terms should include a clear statement that you will disclose those records when served with a valid legal order to do so. Your clients will likely insist, however, that you not make such a disclosure prior to receiving a court order and that you notify them promptly upon receipt of such an order. This type of request by your client is a reasonable one, as it will give them an opportunity to contest the order and to prepare for the impact of disclosure. You should probably make it clear, however, that you do not assume any responsibility for contesting the court order.

Once produced as evidence, records and other information are generally subject to public review. This means that, without special action by the legal authority, the records you have produced can be obtained and reviewed by all parties to the case and by the general public. If there is sensitive material in the records that you produce, you should bring that to the attention of the court, prior to disclosure, and you should request that the sensitive portions of the records be protected from public disclosure. This process is sometimes referred to as placing records *under seal* or, if the amount of material to be protected from disclosure is more limited, the sensitive material can be protected by eliminating (*redacting*) the sensitive portions from the public copies of those records. Even if the legal authority permits you to protect some of the information you deliver from public disclosure, the other parties to your case will have the right to examine all of the material you produce. You may request that those parties be bound by a nondisclosure agreement or order issued by the court in regard to the confidential material, but you must persuade the court to compel that action.

When the records in question contain highly sensitive proprietary material (e.g., trade secrets), you should seek to have that material protected from full disclosure, in some way. The standard approach is to request that the legal authority exempt the material from disclosure based on an argument that full disclosure would result in substantial competitive harm to the disclosing party. If granted, this request would excuse production of the material in question, meaning it would not need to be disclosed. Alternatively, as noted above, you may request that portions of the material be protected from public disclosure by being sealed or by redacting portions of the content.

When trade secrets are involved, requests for protection of the content prior to disclosure are important as a means of ensuring future treatment of the material as a trade secret. The law of trade secrets requires that parties who claim that certain material constitutes a trade secret must take all reasonable actions to protect that material from disclosure. Obviously if compelled to disclose the material under a valid legal order, the party holding the trade secrets must comply. If they do so without at least requesting that the authority excuse production of the material or permit production under seal, there is a chance that parties seeking access to the material in the future may be able to claim that the material is not, in fact, a trade secret. The basis for that future argument would be the fact that the owner of the material did not make all reasonable attempts to protect the material from disclosure in response to the court order.

Placing records in context

All electronic records are potentially evidence that can be used by a legal authority to help resolve a dispute. Not all records are given the same value as evidence, however. In legal proceedings, after evidence is accumulated through the discovery process, the parties attempt to place the evidence in context for the party in charge of resolving the dispute. Put another way, the discovery process provides the court with a great deal of information, but the parties must explain to the court what it all means during the course of the litigation. In countries where an "adversarial" legal process is used, the two sides in the dispute try to highlight the evidence that supports their claims, while rebutting evidence that undermines those claims. In countries where the legal process is not adversarial and is instead one in which the judge or court has the task of identifying the truth, the court takes the lead in evaluating the evidence and determining its factual value.

Electronic records are subject to challenge by the parties in the case and by the court itself. This same process of testing the factual validity of information is applied to all forms of evidence. After evidence has been presented and tested, it is up to the court to determine how much it will rely on that evidence as it attempts to determine the facts associated with the dispute and make decisions to resolve it. Information and records are not automatically accepted as the complete truth by a court. Instead, they are tested and questioned, and they are ultimately given the value as evidence that the court determines they merit.

In the past, electronic records were highly suspect in the eyes of judges and courts. This distrust arose primarily from two causes. Judges did not rely on electronic records in large part because they did not understand them. In addition, judges did not rely on records in electronic form because it was too difficult to ensure that the contents of those records had not been altered.

The bias against electronic records previously held by legal institutions is rapidly disappearing. Today, electronic records are widely accepted by courts and other legal bodies. This widespread acceptance has been largely driven by advances in information technology and the corresponding increase in comfort and familiarity that the legal community has developed with those systems. Acceptance of electronic records as evidence has also been facilitated by new statutes and rules that mandate such acceptance.

In the United States, for example, the Uniform Electronic Transactions Act (UETA) and the Electronic Signatures in Global and National

Commerce Act (E-Sign) legislation illustrate the move toward greater acceptance of the value of electronic records as legal evidence. UETA is a standard statute that has been adopted as law by approximately 18 states. Among other things, UETA recognizes the legal validity of electronic records. It establishes the principle that the legal validity of a record cannot be challenged merely because the record is in electronic form. The E-Sign legislation was enacted at the national level in the United States, and it too establishes the concept that electronic records and traditional hard copy records have equal value as proof of a commercial transaction. Useful information about UETA is available at http://www.bmck.com/ecommerce/uetacomp.htm, and helpful information regarding E-Sign can be found at http://thomas.gov.

These laws do not exempt electronic records from challenge. In legal proceedings, the accuracy and the relevance of electronic records can still be questioned, and if the factual basis for those questions is sufficient, the electronic records can be ignored. These laws simply grant electronic records legal status equal to paper records. They ensure that electronic records cannot be challenged or ignored solely because they are in electronic format.

These laws do not require the use of electronic records under any circumstances. They simply clarify the validity of those records when the parties involved have mutually agreed to use the electronic documentation. In addition, the laws do not mandate any specific type of technology or system for use to create or manage the electronic records. Those decisions are also left to the discretion of the interested parties.

The laws do create certain obligations regarding the form of electronic records. For example, E-Sign permits retention of an electronic version of a canceled check in lieu of the check itself, but it requires that the electronic version contain all of the information from both the front and the back of the hard copy check. E-Sign also requires that electronic records of contracts must be stored in a format that is capable of being retained and accurately reproduced (presumably in hard copy form).

Many different jurisdictions now recognize the legal validity of electronic records and documents. Israel, Tunisia, Venezuela, the United Kingdom, Germany, Argentina, Malaysia, India, Singapore, and South Korea have all enacted laws and rules that recognize the legal validity and enforceability of electronic records. The European Parliament has adopted a policy providing legal validity to electronic records (the Community

Framework for Electronic Signatures, Directive 1999/93/EC), and that directive is now being implemented by the nations of the European Community (EC). The clear global trend is toward wider recognition of the legal validity of electronic records.

Even when electronic records in general are recognized as valid legal evidence, each time specific records or documents are introduced into evidence, that material is questioned and evaluated. The goals of that process are to determine what parts of the material should be given the greatest weight as factual proof and to help the legal authority understand the meaning of the evidence. Today, electronic evidence is increasingly viewed as valid factual proof, but it will always be subject to the same process of questioning and evaluation that is applied to traditional forms of evidence. Documents and records in electronic form should not be denied admissibility solely because they are in electronic form; similarly, they should not be assumed to be truthful or accurate simply because of their form.

Know what records you possess

The basic step in effective electronic records management is to know what records your organization retains at all times. This seems like a simple requirement, but for many organizations, it is not. Step one in effective records management is performance of an audit to identify all forms of electronic records collected and used by an organization.

Many different forms of information qualify as electronic records. Certainly word processing documents, electronic forms, and databases qualify; but so too do less obvious types of electronic information. Electronic mail messages are records. Information associated with electronic commercial transactions (e.g., orders, payment records) qualifies as records. Cookies and other on-line information transfer files are records. So too are Web site–tracking records (e.g., information as to Web page hits and usage and data regarding employee Web use). On-line material that has been cached is another category of electronic records. Encryption keys are a particularly sensitive form of electronic record, as disclosure of that information could compromise the security of communications or transactions.

When assessing what electronic records your organization keeps, think broadly. The most useful approach is to assume that any piece of electronic information that your organization retains is a record that you must account for and manage. When this standard is applied, most organizations find that they possess far more records than they originally believed. Applying this standard, you will recognize that there is information in your organization that you may not actually consider to be a formal record in your normal course of business, yet it is potentially subject to legal discovery. This type of material poses a significant challenge to records management efforts because it is often forgotten as records management policies and practices are developed.

Systems to inventory electronic records should distinguish between records that contain information pertaining to the organization and its activities and those that pertain to other parties (e.g., customers, employees). Records that deal with the organization itself are fully within the control of the organization. Records containing information about other parties (e.g., customer account information, employee personnel files) are not solely controlled by the organization. Generally, the parties who are the subject of those records also have rights regarding the retention, distribution, and use of those records. Inventory and management systems for records should be coordinated with the privacy policies and practices of the organization to ensure that records containing personal information are identified and are processed in a manner consistent with the privacy standards imposed by law and by the organization itself.

Electronic records management is further complicated by the fact that many records are managed and controlled on an outsourced basis. In this setting, another party (a contractor) commonly possesses and controls records that are owned by the client. Effective records management systems and practices must deal adequately with this condition.

Accurate knowledge of all the electronic records retained by your organization is essential in order to protect its legal rights and minimize its risk of legal liability. It is impossible to assert your rights fully or to assess your risks of liability effectively if you do not know what type of information regarding your organization and its activities is accessible in electronic records. If your organization ever seeks to raise a legal claim, it will immediately be called upon to prove the validity of the claim and to substantiate its allegations of the impact (e.g., harm or damages) resulting from the alleged conduct. Only effective knowledge of what you can and cannot prove based on your records will enable your organization to assert those legal rights on a timely basis.

If your organization is the target of a legal claim raised by another, you should, at all costs, try to avoid the situation in which you find out only at the time of a dispute that there is some highly incriminating evidence in your electronic records. There is very little in litigation that is worse than to be surprised by finding, at the last minute, that your own records help to substantiate a legal claim raised by an adversary. Knowledge of your records will help you defend against a legal claim more effectively and it will help your organization exercise better judgment when determining whether continued litigation or negotiated settlement provides the better course of action.

Knowledge of all electronic records that exist in your organization can also help avoid the problem of telling an authority that you do not have certain information, when in fact you do. When responding to a formal order to deliver records, a misstatement about the existence of relevant material can result in fines or even prison time for the individual who made the misstatement. Numerous recent examples of this kind of problem have arisen during litigation and other legal disputes. For instance, White House staff encountered this problem when they advised a congressional committee involved in an investigation that they had provided all relevant documents, and then later discovered that some e-mail messages on the subject had been stored, unknown to the individuals who responded to the congressional inquiry.

Know where the records are and who has access to them

Records management requires that an organization know where all of its records are located. While step one in a records management process is to conduct an inventory of all electronic records, step two must be to identify precisely where all of those records are located. This knowledge is necessary to enable the organization to respond in a timely manner to disclosure orders. For electronic records, authorities often require delivery of both electronic and hard copy versions of those records. The process of preparing those records for transfer and the actual delivery of the records in suitable form is very difficult if an organization does not know in advance where those records are stored.

Effective records management must include the knowledge of who has access to the records and for what purposes. Records are used by authorities to establish facts. The records are valuable only to the extent that their accuracy can be verified. A key part of verification of the accuracy of records is the ability to account for where they have been and who has had access to them. This process is sometimes described as the establishment of a *chain of custody* for records. A chain of custody is basically a list of all of the people who have had access to the records and a description of what they did with those records. By developing this history of control for each record, authorities can do a better job of assessing the likelihood that the record has been altered in some way. This enables them to make a more accurate determination regarding how much evidentiary value should be given to each record. Those records with more solid and dependable chains of custody will generally be treated as more reliable evidence than will those with less documented histories of access and control.

Also important in establishing an effective chain of custody for your electronic records is your ability to demonstrate that those records were physically secure. Your organization should already have physical security measures (e.g., access controls, secure terminals) in place to protect its important records, for business reasons. Those physical security measures will also help to give your records greater legal credibility as evidence, as they will help to demonstrate that the content of the records was not altered.

This census of your organization's electronic records should also include an inventory of the number of copies of those records. Your organization should strive to locate all copies of all electronic records. Admittedly, this is no simple task for some types of records, such as cached content. It is, however, a key step, as in many cases, it will be important to be able to advise a legal authority not only what relevant electronic documents exist, but also how many copies of those documents exist and who had access to them.

The task of locating your electronic records may be complicated by the fact that some of those records may not be managed by employees of the company, but may instead be under the control of a contractor. As discussed previously, many organizations outsource all or part of their computer operations and information technology activities. Under those circumstances, at least some of the organization's electronic records may not be located on the organization's property and may not be within the immediate control of the organization's employees. You must make sure that your contractual arrangements with parties who handle your electronic records provide you with enough enforceable rights to ensure that you will be able to manage those records in a manner that allows you to comply with all applicable legal obligations and enables you to make use of those records effectively to enforce legal rights that your organization possesses.

In many instances, contracts with a provider of outsourced services offer the only means through which an owner of records can effectively exercise control over those records that are created or managed by the service provider. At a minimum, your contract with the service provider should require that party to keep you informed in regard to the complete content and location of your electronic records, even if they are under the control of the service provider. The service provider should be obligated to ensure that those records will be managed in a manner consistent with the policies and procedures your company has implemented for electronic records. Finally, the agreement with the service provider should make it completely clear that those records are your property and must be handled precisely as you specify.

Understand what kind of story your records tell

It is not enough simply to know what electronic records your organization possesses. Each organization must also anticipate how a party outside of the organization would interpret those records. We must assess what the content of those records would mean to an outside party who examines them. We must try to project what type of story those records would tell another party about your organization and its conduct.

If you understand the story that your records tell, you are in a better position to assess your potential liability. You are also better able to evaluate what claims, if any, you may be able to raise. Accurate interpretation of the meaning of your electronic records is essential to help limit your legal liability and to maximize your ability to enforce your rights. Many recent legal cases help to underscore the importance of understanding the type of picture about your organization electronic records will paint if they are disclosed. Microsoft Corporation, for one, presumably learned that lesson the hard way when many of its internal e-mail messages were introduced as evidence to contradict its arguments in the most recent antitrust case brought by the U.S. government against the company.

To understand what your records will mean to others, your organization must be aware of the content of all its electronic records. This requires some system to create summaries of the records and to index them. The summaries must be accurate and the indexes must enable you to find the records based on standard search parameters (e.g., subject, parties involved, type of record, type of transaction). Of particular importance is the need to update these indices and summaries continuously to ensure that they are always accurate and current.

Remember that you are free to select those records that you choose to retain, unless there are specific legal requirements that set a specific period of time for retention of a certain type of record or until you become aware of the possibility of some form of litigation. If the law requires that you retain certain records for a set period of time, you must hold those records for the required period. Tax records provide one common example of a specific type of record that must be retained for a set period of time.

If you become aware that there may possibly be litigation involving a specific topic, you must then retain all records relevant to that topic. Destruction or alteration of records governed by specific retention requirements or in anticipation of litigation is illegal and can result in fines or imprisonment. Such alteration of documents when there is reason to believe that a legal dispute may exist can be considered to be destruction of evidence or tampering with evidence, and both of those activities are illegal. The time to make decisions about retention of records is far in advance of any type of legal dispute or controversy. Those decisions should be handled in the form of a clear and consistently applied records retention and management system, based on carefully planned policies and practices.

Recognize that once your organization retains certain records, it will be responsible for acting appropriately on the content of those records. This point is perhaps best illustrated with regard to employer monitoring of employee use of the company computer system. Many employers now monitor their employees' use of e-mail, word processing, and the Web. That monitoring is generally conducted for sound legal and commercial reasons. Employers want to try to make sure that their employees do not waste the computing resources of the company, they want to promote data security, and they want to try to minimize the chances that they will face legal liability as a result of improper employee conduct (e.g., harassment or securities law violations).

Companies that conduct employee monitoring must, however, recognize that they are creating electronic records. When they track e-mail, Web use, on-line transactions, or even the keystrokes of employees at their computer keyboards, companies will be held accountable for knowledge of the content of those activities. If, for instance, an organization tracks the Web surfing activities of its employees, and one of those employees commonly visits Web sites presenting racially based hate content, the employer will be held responsible for knowledge of that conduct, and the law in some jurisdictions, including the United States, may require the employer to take action against the employee to stop the conduct. Once you create a record, electronic or otherwise, you have verified that you have access to the information contained in that record. As illustrated above, in some instances, mere knowledge of certain activities imposes a legal requirement to take action. Your electronic records thus tell other people stories about your company and its actions, and their mere existence may force your company to take certain remedial actions. If your organization fails to take remedial action after its records indicate that it has knowledge of the illegal conduct, the organization may also be held legally liable for the conduct.

In some instances, the electronic records may provide evidence of illegal conduct by the organization itself. For example, assume a company with a substantial international operation discovers in its e-mail records that information covered by U.S. export control regulations has been distributed to foreign parties without the required license from the U.S. Department of Commerce. Upon such a discovery, the company has a legal obligation to report the apparent violation to the proper law enforcement authorities. The company must turn itself in based on the information it discovered in its records.

Generally, penalties assessed against organizations that take this type of remedial action when they become aware of the violation are substantially less severe than those assessed against organizations that do not figure out that they have a problem until the government authorities initiate an investigation. Note that the harshest penalties are generally assessed against organizations that discover their violations but take no remedial actions, hoping instead that no one will notice. The basic rule of thumb is that if your organization is doing something inappropriate, assume someone will eventually find out about it. Also assume that your organization will most likely be better off in the long run if it finds the problem before the authorities do, and if it makes the initial disclosure to the authorities.

Given the many challenges associated with managing electronic records, organizations should think carefully before they retain additional records. Before your organization elects to create another form of electronic record, the potential legal consequences of that action should be reviewed. Evaluate what type of information the proposed new records will contain, and consider whether possession of that new information triggers any new legal obligations. Also consider what potential commercial consequences would be associated with any mandatory disclosure of the new information that might be ordered by a court or other legal body during litigation. Ask your organization the question: If a neutral party saw the content of those new records, what kind of reaction would that party have?

Implement policies and practices to manage those records

After we identify and evaluate the electronic records of our organization, we must develop and implement policies and procedures to manage those records. The policies identify records management objectives and the practices provide the operational means through which those objectives are to be met. The policies and practices should be expressed to the organization in a statement or manual. That document should address several key topics, including the following.

The statement should describe the different categories of records that are retained by the organization. This list can be quite long as it will include easily identified material such as word processing documents and e-mail, but it will also include somewhat more obscure electronic records such as cached on-line content, stored cookies, Web site use information, and employee computer use data. For each of these different categories of electronic records, specific management practices should be described. Those practices should specify: (1) how many copies of the records will be retained, (2) where the records will be stored, (3) who will have access to the records, (4) how long the records will be retained, (5) whether hard copies of the records (or hard copies of lists of the records) will be retained, and (6) how the records will be destroyed and who will destroy them.

An individual employee (or position) in the organization should be specifically identified as having responsibility for management of each type of record. It is reasonable that many of the different categories of electronic records will be under the direction of the same individual, but there should definitely be a single individual identified as the senior corporate staff member who has company-wide responsibility as guardian or custodian of all the organization's electronic records. Different organizations take different approaches to this issue, as some make the chief information officer the records custodian, while others have that role played by a senior legal, administrative, or financial officer. Some enterprises have a chief privacy or security officer, and that individual is sometimes tasked with the role of electronic records guardian. Regardless of which position fills that role, the important point is that the individual playing that part has a background that enables him or her to effectively manage the interesting blend of technology, business operations, and legal/financial topics that affect records management. The document custodian is likely to be the individual who will be required to attest personally to the policies and practices of the organization, and that individual could face personal legal liability if illegal conduct is involved.

Although a custodian for each category of electronic record should be identified, all employees should be made aware of the records policies and procedures and should be ordered to comply with those standards. All employees should be obligated to follow the directions of the records custodians as to management and use of the records. Violation of policies and procedures applicable to the records should be specifically identified as grounds for dismissal from employment.

Locations for storage of the records should be specified. Instructions should also be provided to the guardians of the records specifying what security measures will be applied to the stored records. Security instructions for the records are necessary to provide a chain of custody to verify that the records have not been tampered with or otherwise altered. Security instructions should include guidance regarding which individuals will have access to the different types of records and for what purposes.

Guidelines for records management should also deal with the issue of unintentional loss of, or damage to, those records. The guidelines should include notice procedures in the event of such loss or damage and instructions as to replacement or reconstruction efforts. One option to be considered is insurance for some of the most critical records of the organization, as some insurance carriers now provide such coverage.

Records requiring special management procedures should be identified. Records that are particularly sensitive for the organization (e.g., those that contain trade secrets) and records that contain information about other parties (e.g., customers, suppliers, employees, investors) should be treated with special care. Records such as medical or financial records that are subject to special legal requirements should be identified. The specific instructions for their use, storage, and destruction should be clearly defined, and should be consistent with the special requirements imposed by law for those records.

The records manual should require periodic audits of electronic records. Those audits should be designed to verify that the established records management policies and practices are, in fact, being followed. The audits should also be structured so that they will identify new forms of electronic records that the organization now retains. In some cases, changes in technologies or business practices may result in the unintentional creation and retention of new forms of electronic records. Periodic audits will enable organizations to identify those new forms of records and to modify the policies and practices, as necessary, to accommodate the records.

The manual governing management of electronic records should be reviewed and updated on a regular basis. It should be updated to reflect changes in the operations of the business. It should also be updated to accommodate the different rules of the many different jurisdictions in which the company does business. Electronic records management must reflect the legal rights and obligations imposed by different states and nations that have jurisdiction over the company. This part of the management process can be particularly challenging for companies that conduct business globally. Multijurisdictional compliance is an essential element of electronic records management and the best way to ensure such compliance is to make sure that the policies and practices are updated frequently enough to reflect the often rapid changes in the number of jurisdictions in which the company conducts business.

All employees and contractors of the organization should be trained as to the records management policies and requirements. Training should be provided for all new employees, and periodic retraining should be provided on a regular basis. When the policies and procedures are modified, all employees and contractors should be made aware of the substance of those changes.

The statement should also clearly indicate what action should take place in the event that a record is found to contain evidence of illegal or inappropriate conduct. Under those circumstances, the statement should require that any employee who discovers such evidence must report the finding to an appropriate member of the senior management team (e.g., the records custodian). The manual should require that senior management, when made aware of the problem, take all appropriate actions required by law and company policy to remedy the violation (including disclosure to law enforcement authorities, if necessary).

As business decisions regarding creation and retention of new forms of electronic records are considered, the potential legal implications of retaining those additional records should be evaluated before final decisions are made. The manual should prescribe a process through which the potential legal impact of retention of additional electronic records is factored into the business and technical decision about whether or not to establish and retain those additional records.

The manual should also provide guidance as to coordination of electronic records management policies and procedures with those applied to records in other forms. Some organizations may choose to deal with both traditional and electronic records in a single statement, but if different systems are to be applied, it is essential that guidance be provided to ensure coordination of the systems and consistent application. You must avoid, for instance, a situation in which a hard copy version of record is treated one way, and an electronic version of the same record is managed in an inconsistent manner.

General legal protection for electronic records

Some jurisdictions provide broad legal protection for the security of certain electronic records. In the United States, for example, the Electronic Communications Privacy Act (ECPA) prohibits unauthorized access to electronically stored communications (e.g., stored e-mail messages). The scope of this statute expands as more diverse forms of communications are commonly stored in electronic format (e.g., voice conversations, facsimile transmissions). When stored communications have been accessed without permission, the party accessing those records may be liable under the ECPA. The ECPA also prohibits the party storing those communications records from disclosing those records unless the disclosure is necessary under a legal order or it is necessary for the provision of communications services or to protect the disclosing party's property rights. Penalties for ECPA violations include mandatory monetary payments and court-ordered remedial actions.

In the United States, the Computer Fraud and Abuse Act (CFAA) prohibits unauthorized access to electronic records stored on computers that are used by, or for the benefit of, the U.S. government. The CFAA also prohibits such unauthorized access when financial institutions retain the records involved. The CFAA is also applicable when there is unauthorized access to computers or computer systems involving interstate computer use, even if those computers are not handling financial or government-related data. Penalties for CFAA violations include mandatory monetary payments and court-ordered remedial actions. The CFAA also provides for criminal penalties (including prison time) for violations under some circumstances.

For both the ECPA and the CFAA, legal liability is asserted against the parties who access the electronic records in question without authorization. Liability can also be assigned to the party who manages those stored records if that party assisted in the unauthorized access. That liability may be present even if the party who stored the records did not intend to assist in the disclosure. Managers of stored electronic records that fall within the scope of the ECPA or the CFAA should, accordingly, take special precautions to avoid facilitating, in any way, unauthorized access to those records. Compliance challenges are becoming increasingly difficult as the scope of electronic content increases.

Financial, health, and medical records

The law requires special security management for certain types of electronic records. A good current example of such specific regulation is the developing federal standards in the United States for management of health and medical records in electronic form. These types of records receive special attention at law because policymakers consider them to be particularly sensitive and important.

The HIPAA was enacted by the U.S. Congress in 1996 and the regulations implementing the statute are currently being developed. It is expected that development of the HIPAA regulations will be completed during 2001, and if that schedule is met, full compliance with all of those regulations will be required early in 2003. Useful information regarding the HIPAA is available from several on-line sources, including http://www.hipaadvisory.com and http://aspe.hhs.gov/admnsimp/.

The HIPAA has three main objectives: (1) standardize the format for electronic patient health, administrative, and financial data; (2) develop unique identifiers for individuals, employers, health care providers, and health plans to use in the health care delivery and management process; and (3) set security standards to protect the integrity and privacy of individually identifiable health information. As to electronic security, the HIPAA adopts specific standards to be applied to the physical storage, maintenance, transmission, and access to health information that can be matched to a specific person.

Certain financial records are now the subject of security guidelines imposed by the U.S. government as a consequence of financial deregulation legislation enacted in 1999 (legislation commonly referred to as the Gramm-Leach-Bliley Act). Those federal guidelines, effective in 2001, place security obligations on institutions that collect personally identifiable information from individuals who have obtained, or are obtaining, a financial product or service primarily for personal, family or household use. The guidelines apply both to electronic and hard copy consumer financial records. Useful information on the guidelines is available at a variety of sources, including http://www.privacyheadquarters.com.

The Gramm-Leach-Bliley guidelines basically encourage parties who handle nonpublic personal information regarding financial activities of consumers to implement security measures to protect the privacy of the content of those records and to provide notice to consumers so that they understand how those records will be used and protected. These provisions are presented in the form of guidelines, not formal regulations. This means that compliance is not technically mandatory. Wise businesses will, however, move to meet the guidelines, as they are likely to provide the basis for future formal regulations; thus prompt compliance will probably reduce compliance costs in the long run. Prompt compliance will also provide businesses with some protection from potential legal liability in private, civil lawsuits brought by customers. Compliance is also likely to be necessary for commercial competitive reasons, in order to make your organization's products and services as attractive as those offered by competitors who meet the guidelines.

Mandatory records

The laws of many jurisdictions require the collection and retention of some records. One of the most common examples of this type of mandatory record retention in the United States is the required collection of transaction data associated with sales of goods. The transaction data is collected so that appropriate sales and use taxes will be paid to those states entitled to receive the revenues. Although the issue of sales and use taxes applicable to sales made using the Internet is still the subject of substantial debate and disagreement in the United States, current law still requires payment of appropriate sales or use taxes when the seller involved in the transaction has a physical presence (*nexus*) in the state in which the buyer resides. Failure to make the appropriate tax payments to the proper government authority can result in legal penalties for both the buyer and the seller.

Electronic sales of goods in the United States that trigger a sales or use tax obligation must be properly documented. Records that identify the relevant state, the buyer, the amount of the purchase, and other information necessary to facilitate timely payment of the associated tax amounts must, accordingly, be retained by the seller. For electronic sales, most of these relevant sales records will likely be electronic documents. This record management obligation will become increasingly important as electronic sales transactions increase in frequency and in economic value.

Public records

While laws protect some electronic records from public disclosure, other records are made part of the public record through mandatory disclosure. One example of required disclosure consists of records documenting government actions. In the United States, at both the federal and state level, records that implement and support government decisions and actions are made available to the public through laws such as the Freedom of Information Act (FOIA). FOIA and similar laws permit members of the public to request that government records be disclosed for open review. FOIA and similar laws apply to electronic records as well as traditional hard-copy documents. As more and more government actions are taken in electronic format and are based on input provided in electronic form (e.g., e-mail), an increasing number of electronic documents will be subject to the disclosure provisions of FOIA.

Another example of mandatory disclosure of records is provided by securities laws in the United States and other nations. These laws require that private companies that sell equity ownership to the public make certain basic corporate records and information available to the public, so that shareholders and potential shareholders have access to information relevant to their decision about whether they will invest or continue to hold their investment in the company. In the United States, these securities disclosure rules are enforced by the Securities and Exchange Commission (SEC), at the federal level, and by state regulators, as well. Mandatory securities law disclosures are now commonly processed through electronic systems (e.g., the SEC's EDGAR reporting system). Many of these mandatory securities law disclosures involve collection and distribution of electronic financial records of the publicly traded companies.

Mandatory disclosure of government records under FOIA and "sunshine" acts, and financial reporting disclosures under securities laws, are two of the most common examples of government-imposed records release requirements applicable to both traditional hard-copy and electronic records. In both instances, policymakers concluded that the general benefits associated with more widespread access to the information contained in those records outweighed any negative impact associated with the cost of disclosure and the loss of privacy or secrecy for their content. As more documents and records are retained in electronic form, these mandatory disclosure rules will have a growing impact on electronic records.

Other electronic records

Many organizations now retain certain electronic records for reasons other than legal obligation. For example, certain electronic records associated with commercial transactions or financial activities are commonly retained to facilitate audit and accounting activities. In these instances, investors and lenders encourage companies to establish and retain electronic records regarding financial and commercial transactions and activities to help make the financial review and auditing functions more accurate and more visible. Creation of an electronic audit trail can help to improve the operations of enterprises and enhance the ability of company stakeholders to exercise their rights.

In other cases, enterprises choose to retain electronic records to help support a variety of basic commercial functions (e.g., insurance). Organizations with thorough electronic records systems can sometimes earn reduced insurance premiums or other business advantages as a result of those systems. The rationale behind those benefits is that the records systems enhance operating efficiency or reduce operating risks at a level sufficient to merit the favorable treatment.

It is highly likely that these nonmandatory commercial reasons for retaining electronic records will become yet more important in the future. If this happens, the incentives for expanding the scope of electronic record-keeping will grow. As we have seen throughout this chapter, however, more electronic records means more records management challenges for businesses.

International aspects

Electronic records are used as evidence in jurisdictions around the world. They are relevant in litigation between private parties and in cases initiated by governments. As organizations expand the scope of their computer networks, they find that their electronic documents are located in an increasing number of different legal jurisdictions. Those records thus become subject to the rules governing evidence and disclosure of special forms of information in many different nations, regions, and localities. The global scope of those operations complicates the electronic records management process.

Various countries have developed legal requirements for the protection of certain forms of particularly sensitive personal records. The electronic records most frequently protected by this type of regulation include records that contain information pertaining to specifically identifiable individual people (personal information), including financial and health/medical records. U.S. legal action regarding these categories of electronic records was discussed above. The EC has implemented sweeping information privacy requirements to protect personal information from disclosure, and those requirements are addressed more fully in the discussion of information privacy later in this book. Regulatory action has been taken by countries such as Australia, which makes unauthorized access to electronic records of financial institutions a criminal offense subject to penalties including prison terms of up to two years. Some jurisdictions have elected to deal with the issue of protection of electronic records through laws that prohibit unauthorized access to computers and their content. Those laws are reviewed in the discussion of control over computer system access later in this book.

In addition to requirements protecting certain records from disclosure, in some instances there are legal obligations to release those records to authorities. Some countries have made certain electronic records the target of specific retention or disclosure requirements. For example, a controversy developed in the United Kingdom as a result of the enactment of the Regulation Investigatory Powers Act (RIP) in 2000 (additional information regarding the RIP is available at http://www.legislation.hmso.gov.uk/acts/acts2000/10000023.htm). RIP was controversial for several reasons, including the fact that it limits the ability of a person who is required by the U.K. government to surrender encryption keys from disclosing to any other party the fact that the keys have been surrendered. The fear is that this could lead to situations in which an organization's encrypted communications can be accessed by the U.K. government, yet the target organization is unaware that the communications have been compromised, as the party responsible for the disclosure to the government is under a legal obligation not to disclose the fact that the key had been surrendered.

Various countries require that certain electronic records be retained or, in some cases, that they be provided to the government. Countries such as Cuba, Vietnam, and China, for instance, require that information regarding ISP accounts and Web sites be registered with the government, as a matter of standard operations. Not surprisingly, many organizations and individuals are wary of requirements for electronic records retention or disclosure imposed by governments. Those legal requirements provide a major source of information for governments, and they afford a level of visibility into the operations and activities of enterprises and individuals that is not entirely justified. Nevertheless, those electronic records requirements are not merely isolated incidents, and it is likely that an increasing number of governments will move to gain access to additional records of service providers and network users.

Organizations are becoming increasingly sensitive to the fear that many different government authorities around the world may begin to require disclosure of certain highly secure electronic records. As noted previously, one of the most sensitive forms of electronic records is an encryption key, because its disclosure can compromise the security of communications. Other highly sensitive types of electronic records include the registration and identification records of the certification authorities that are being developed to help foster security in electronic commerce. Access to these records, by governments or by private parties, is cause for significant concern to proponents of electronic commerce. If your organization is involved in any of these operations relying on highly sensitive records, you should recognize that your records will likely be the prize targets of governments and private organizations around the world.

Just as you develop your systems and operations to protect those records from unauthorized users, so too should you act to make sure that they can be protected, as much as possible, from widespread compulsory disclosure. One way to begin to accomplish this goal is to monitor carefully the potential legal disclosure obligations in all of the jurisdictions in which you conduct business. You should try to influence the development of those rules whenever possible. If the disclosure rules in certain jurisdictions are particularly burdensome, it may well be wise to consider adjusting your business operations to minimize your exposure to those jurisdictions by minimizing your physical presence and the scope of your business activities in those places.

As companies expand their international operations, they will be required to comply with an array of rules associated with electronic records applied by the different countries in which they conduct business. Those rules will require the protection, retention, and disclosure of certain types of records. Electronic records management systems implemented by businesses should, therefore, recognize the international scope of business activities and should be designed to facilitate compliance with many different records retention and use rules. Even with that attention, however, coordinating multijurisdictional compliance for electronic records management is no small task.

Appendix 2A: Electronic records management checklist

The following basic steps are essential to effective management of an organization's electronic records.

Inventory records

The first step in effective management of electronic records is to complete a thorough inventory of the records retained by the organization. The inventory should include identification of the following information: (1) the form or type of records retained (e.g., e-mail message, electronic data interchange material, cookie, Web usage–monitoring data); (2) location of the records; (3) number of copies (both electronic and hard copy); (4) summary of content of records; (5) people with access to the records; (6) description of permissible uses of the records by authorized users; (7) records that contain information about parties other than the company itself; (8) records subject to special legal requirements (e.g., employee records, customer records, trade secrets); and (9) records of the organization that are created, maintained, or stored by other parties (e.g., contractors, business partners).

Designate records custodian

The organization should designate custodians for each category of electronic record. A member of the senior management team of the organization should be identified as the custodian or guardian for all of the electronic records of the organization. That individual should have authority over all other records custodians. The records custodian should have the authority to enforce all records management policies and practices, and should speak for the organization's records. The custodian should be the party to provide all attestations and certifications associated with the records.

Create records management policies and procedures

All records management policies and procedures should be clearly expressed in written form in a statement or manual. The manual should be provided to all employees and contractors working with the company. The manual should be updated on a regular basis, and it should clearly indicate that failure to comply with the policies and practices defined in the manual is a basis for termination of employment.

Train/educate employees and contractors

Training sessions for all employees and contractors should be provided to explain the requirements imposed by the records management manual. Participation in those training sessions should be mandatory. Additional briefings should be held each time the manual is updated. All employees and contractors should be required to sign written statements verifying that they have reviewed the manual, been briefed on its contents, and agree to abide by its terms.

Appendix 2B: Sample topics for electronic records policies and practices manual

An organization's electronic records policies and practices should be clearly expressed in a written statement or manual provided to all employees and contractors. That manual should be reviewed and updated regularly, and it must be effectively enforced (once you create it, the law will expect you to comply fully with all its terms). The manual should include the following topics.

List of categories of records

The manual should describe all of the different types of electronic records retained by the company. Although this list may eventually become a long one, for some organizations, there is value in attempting to identify a comprehensive list. The list should be updated frequently so that it remains accurate. The categories should highlight records that are subject to specific legal requirements (e.g., securities information, trade secrets), those that contain information about parties other than the organization itself (e.g., customer records), those that contain particularly sensitive content (e.g., employee records), and those that are in the possession or under the control of a party other than the organization (e.g., a contractor).

Description of requirements for each category of records

For each electronic record category, the manual should describe the following management practices: (1) number of copies retained (electronic and hard copy), (2) location of storage of all copies, (3) parties with access to the records and for what purposes, (4) period of time the records will be retained, and (5) how the record will be handled at the end of retention term (e.g., destroyed, transferred to another party), and who will be responsible for that action.

Identification of custodians

The manual should identify the positions of the people who will serve as the custodians of the organization's electronic records. The manual should indicate that records custodians have the authority to enforce all records requirements established by the manual, that they are the only individuals authorized to speak for the organization on topics pertaining to its records, and that they are the only individuals authorized to release records to parties outside of the organization. All requests for records or inquiries regarding records made by parties outside of the organization should be directed immediately to the appropriate records custodian.

Obligations of employees/contractors

The manual should identify the following obligations for all employees and contractors: (1) review the manual, verify understanding of the manual's content, and abide by the policies and practices defined in the manual; (2) follow the directions of the records custodians as to use of the records; (3) report all violations of the terms of the manual to a records custodian; and (4) avoid disclosing or describing any of the organization's records to a party outside of the organization without the prior written approval of the appropriate records custodian. The manual should clearly indicate to employees that failure to abide by the terms presented in the manual is cause for termination of employment. A procedure through which employees can raise questions regarding the electronic records policies and practices should be established in the manual.

New records and audits

The manual should provide a system through which the records custodians can participate in the decision-making process associated with the creation of new categories of electronic records. As business and technological decisions which may result in the creation of new forms of electronic documentation are considered, the records guardians should have the ability to participate in those deliberations to provide insight as to the potential consequences of creation and retention of new types of electronic records. The manual should also define a procedure for periodic audits of the records management system to ensure compliance with the requirements of the manual and to ensure that the terms of the manual remain comprehensive.

Coordination with other policies and practices of the organization

The manual should provide a system to coordinate the electronic records management policies and practices with other important requirements imposed by the organization. Electronic records policies and practices should, for example, be closely coordinated with the organization's approach to its hard copy records. Similar coordination with information privacy policies, computer system use policies, and employment policies is also required.

Appendix 2C: HIPPA records compliance

The HIPAA imposes certain obligations as to the security of some health and medical records. Those requirements apply to all health care providers, which includes all organizations that participate in the delivery of health care services (e.g., medical service providers, employers, educational institutions, insurance companies) and the delivery of services ancillary to basic health care (e.g., billing organizations/payment processors, developers or operators of software/information systems supporting health care delivery). The following outline provides a basic action plan for HIPAA compliance.

Identify relevant records

Organizations involved in health care delivery should promptly identify all records that they retain or process which fall within the scope of HIPAA. Basically, those records include all nonpublic health information that can be attributed to a specific individual. This is a broad definition of confidential material; thus many records are within the scope of the HIPAA.

Evaluate current security compliance

Each organization that retains or processes electronic records within the scope of HIPAA should evaluate the current state of their security measures for those records. This review should examine both the measures applied to stored records and those applied to electronic transactions that process the relevant information. The current security measures should be compared with the standards for storage security and transaction security now under development by the American National Standards Institute, in conjunction with the U.S. Department of Health and Human Services.

Implement security measures

If current security measures do not comply with the standards under development, those measures should be upgraded to permit compliance. Security standards more rigorous than the mandatory standards are, of course, acceptable.

Implement notice procedures

An important aspect of the HIPAA is the notice provisions. The proposed rules give individuals the right to receive written notice describing the information security and privacy practices of the organizations that handle their personal health information. It also gives those individuals the right to review their information and to amend it. Organizations that handle such records must be sure to institute systems that will ensure the notice and review obligations are satisfactorily met.

Manage disclosures/use of records

The proposed rules limit the disclosure of the personal information to the minimum amount necessary to provide the requested health care services. Prior written authorization from the individual is required before the information can be used for purposes other than treatment, health care operations, or payment for health care services. This means that information management systems must be in place to prevent distribution or use of the personal information for purposes prohibited by the HIPAA, and to obtain the requisite written authorizations when needed.

Create audit trail

HIPAA requires that an effective audit trail be created for all health information disclosures. This means that if an organization intends to disclose personal health information, a verifiable record of those disclosures and the safeguards applied to ensure HIPAA compliance must be established. That audit trail must be sufficient to permit a neutral party to verify HIPAA compliance as to the disclosed material.

Train/educate employees and contractors

Effective HIPAA compliance will require that many different employees and contractors in your organization understand the requirements of the act and the associated regulations. Training programs are an essential part of that compliance.

Provide for system upgrades

As security technology improves, it is likely that HIPAA standards may eventually be revised. With that in mind, it makes sense to structure the security upgrades you make to meet the developing HIPAA standards in a manner which facilitates future upgrades that can be made promptly and efficiently as new security technologies and systems develop.

Appendix 2D: Gramm-Leach-Bliley Act financial security guidelines action plan

The guidelines adopted to implement the records security aspects of the Gramm-Leach-Bliley Act apply to disclosure of nonpublic information about individuals who have obtained (or are obtaining) a financial product (or service) from a financial institution primarily for personal, family, or household purposes. The guidelines encouraged all organizations that collected or used such information to disclose privacy and security policies and practices applicable to that information to their customers by July 1, 2001. That notice should have been provided clearly and conspicuously in written documents and through electronic media (e.g., at Web sites). The following actions provide a reasonable approach to compliance with those guidelines.

Identify relevant records

Each organization that collects or uses the nonpublic, personal financial information addressed by the act should identify all of the records it handles that fall within the scope of the act.

Evaluate current security risks

After identifying the relevant records, the organization should assess the security risks associated with the collection and use of that information. It should then consider whether current security measures are adequate to protect the privacy of that information given the expected level of threat.

Establish security policies and practices

Security policies and practices should be developed (or enhanced if they are already in place) as appropriate, given the security risk assessment made with regard to the personal financial information.

Create audit trail

Security policies and practices should be designed so that compliance with those requirements and the effectiveness of the requirements can be verified through an audit by an independent party. The security system and procedures should be structured in a manner such that an audit trail is provided.

Train/educate employees and contractors

Compliance with the guidelines requires understanding of the guidelines and of the policies and practices implemented to comply with the guidelines. Employees and contractors involved with the processing or use of personal financial information of consumers should participate in mandatory training sessions.

Provide for system upgrades

As security technology and systems become more sophisticated, these security guidelines are likely to be modified to reflect those advances. Future costs and frustrations can be minimized if the security measures implemented today to meet the current guidelines are structured so that future upgrades can be introduced quickly and efficiently. Compliance with today's security guidelines should be performed with an eye toward future security capabilities.

3

Preventing Unauthorized Access

A major aspect of security in the digital marketplace is the prevention of unauthorized access to computer and telecommunications systems. Unauthorized access events can be the basis for legal liability attributed both to the intruder and to the operator of the compromised system. Governments and private parties have legal rights that can be enforced against parties engaged in unauthorized computer network access. Third parties who are harmed by the unauthorized access (e.g., owners of intellectual property) also have rights that may be enforced against the party initiating the unauthorized access, the owner of the computer system, and the operator of that system.

Under some circumstances, certain forms of access by outside parties are required by law. Surveillance conducted by law enforcement or national security authorities, subject to appropriate legal authorizations, provides one example. In those instances, computer system operators must understand when such access is required by law and when it is not. Failure to facilitate access when required by law can lead to legal liability for the system owner and operator (e.g., criminal penalties for failure to comply with court orders). Yet, grant of access when not required by law, can lead to liability for the system owner and operator from third parties who may be harmed by that access (e.g., system users who have rights of privacy or trade secrets violated by the access).

Different forms of unauthorized access

From a legal perspective, there are three key forms of unauthorized computer access: (1) unauthorized access to the computer system and equipment, (2) unauthorized access to the data stored and processed by the system, and (3) unauthorized access to access codes or other means of gaining access to the computer system. The laws in many different jurisdictions address each of these different aspects of the unauthorized computer use issue. Legal remedies for unauthorized computer system access include both criminal and civil (or private) legal claims.

Criminal penalties

Criminal laws are the laws that are enforced by the government against private parties. They commonly carry penalties of fines (i.e., mandatory monetary payments) and imprisonment. In criminal cases, the government is the party that raises the complaint (i.e., prosecutes the case) and the defendant must answer to those charges. The laws that form the basis for criminal court cases are different from those applied in civil lawsuits.

Criminal penalties against unauthorized system access

The United States and many different nations now have criminal laws that specifically prohibit unauthorized access to computer equipment and computer systems. These laws prohibit unauthorized access, no matter what action the party may undertake after gaining illegal access to the system. These laws do not generally require a showing of information theft or damage. Instead, they are sanctions against the act of unauthorized access itself. Unauthorized access may consist of an intrusion involving nothing more than access to the computer system for the simple purpose of proving that such access was possible. Unauthorized access also generally includes situations in which an authorized user of a computer system exceeds his or her authority, and thus accesses computers or their content in a manner that is beyond the scope of the authority possessed by that user.

The computer systems protected by these laws are diverse. Laws against unauthorized access generally protect individual computers, networks of computers, and telecommunications systems. Many of these laws were originally created to protect telecommunications networks (e.g., the voice telephone network, cellular phone systems) from misuse. Most of these laws are drafted broadly enough to provide protection for a virtually limitless set of different types of computer systems. While they obviously protect the telecommunications networks, ISP systems, and private computer networks, the laws also offer coverage for other forms of computers. For example, intelligent appliances with computing capability would come within the scope of computer access laws, as would other types of products (e.g., automobiles) to the extent that they contain computing devices (e.g., intelligent microprocessors).

In the United States, the CFAA makes it a criminal violation to engage in unauthorized access of computer systems that process certain forms of data (e.g., financial data and government data). The CFAA also prohibits the use of interstate or foreign communications systems to gain unauthorized access to a computer or a computer network, no matter what content that network may contain. The penalties associated with the CFAA include fines and prison terms of up to 10 years.

Other countries around the world also provide legal protection for computers and computer networks in their criminal laws. Portugal and Singapore prohibit unauthorized computer access in their criminal laws. Unauthorized access to computers or computer networks in Portugal can be penalized by fines and jail terms of up to three years. In Singapore, criminal penalties include fines and prison terms of up to seven years.

Japan and The Netherlands also impose criminal penalties for unauthorized computer access. In Japan, such access is subject to fines and prison terms of up to one year. In The Netherlands, unauthorized computer access is deemed to be a breach of the computer peace and is punishable by a jail term of up to six months, along with fines. The laws in Japan and The Netherlands require that the computer that has been compromised be one that was subject to some form of restricted use or security system. In those countries, if an unauthorized user accesses an unprotected computer or network, that access will not necessarily be a violation of the law.

The requirement that some form of security system or use restrictions be in place is an important point for computer system operators to keep in mind. As the criminal laws in Japan and The Netherlands demonstrate, some jurisdictions require that the computer system operator take reasonable steps to prevent unauthorized access before those laws will prosecute an unauthorized user. If effective security measures and use restrictions are not in place, the notion is that the law should not necessarily provide remedies to compensate for the absence of those protective measures. The existence of laws prohibiting unauthorized computer access alone, will not necessarily protect computer operators if they do not make all reasonable efforts to help themselves through the application of prudent security systems. The lesson here is, while the law can provide an important tool to help protect networks from unauthorized access, owners and operators of those networks must also do the best they can to provide their own protection for those systems, through prudent use of security technologies and systems. Do not rely on the law alone for computer security.

Finland and Luxembourg also prohibit unauthorized computer access. In Finland, that unauthorized access is considered to be data trespass, for legal purposes. It is punishable in Finland by fines and jail terms of up to one year. In Luxembourg, fines are also permitted and jail time can range from two months to one year. The law in Finland also provides an example of an approach adopted by several jurisdictions in the way it treats access attempts. Finland even applies its penalties for unsuccessful attempts at unauthorized access.

Criminal statutes in France and Germany also prohibit unauthorized computer access. Fraudulent access to a data processing system in France is subject to fines and jail terms of up to one year. In Germany, unauthorized access can be penalized by up to three years in prison. Even unsuccessful attempts to gain access are violations of German law. It is important to be aware of the fact that several countries adopt an approach similar to Germany's, applying unauthorized access penalties to all attempts to gain access, not only the ones that are successful.

Belgium, Italy, and Israel also prohibit unauthorized computer access. In Belgium, penalties include fines and prison terms of up to two years. Unsuccessful attempts to gain access are also subject to the criminal penalties. Italy provides for fines and jail terms of up to three years for unauthorized system access, as does Israel. Even jurisdictions that do not have criminal statutes that specifically prohibit unauthorized computer access commonly find ways to prosecute parties who engage in such conduct. The most common approach under those circumstances is to find ways to prosecute the defendant under the already existing, traditional criminal laws.

For example, while criminal statutes in Norway do not specifically address computer system access, they do address issues such as unauthorized access to documents, and deliberate attempts to damage infrastructure (e.g., telecommunications and broadcasting systems). In many instances, those statutes would be sufficient to permit criminal prosecution of parties in the context of unauthorized computer use. Although there have been highly publicized examples of situations in which authorities claimed that their traditional laws were inadequate to prosecute cases involving computer misuse (e.g., the Philippine government's difficulties during the "I Love You" virus), it appears that most jurisdictions have criminal and civil laws that are adequate to permit some form of legal action against parties who misuse computer systems. It is seemingly a rare situation in which a government has no legal recourse against computer abuse. Even so, the more active movement in many countries and regions to review their laws in order to determine if they provide adequate coverage for computer access and use problems, and modify or add to them as appropriate, is prudent.

Criminal penalties against unauthorized data access

A second approach taken by some jurisdictions on the issue of controlling unauthorized computer system use is the establishment of criminal penalties against unauthorized data access. While some criminal laws focus on access to the computer equipment, others focus on access to the data stored, communicated, or processed by that equipment. Those laws penalize access or attempted access to material stored or communicated on computers and computer networks, without the appropriate permission of the owner of the computers. In the United States, for instance, unauthorized data access can be punished with fines or prison terms of up to 10 years (or longer, if the offender has a prior record of computer abuse).

Denmark, Austria, and Australia prohibit unauthorized data access. In Denmark, such unauthorized access to computer content is subject to fines and prison terms of up to six months. Criminal laws in Austria provide for fines for unauthorized data access. In Australia, penalties include fines and prison terms of up to two years.

Germany, Greece, and Ireland also provide for criminal penalties in the event of unauthorized data access. Germany considers such access to be data espionage, and penalizes that conduct with fines and jail terms of up to three years. Greece imposes fines and prison terms of up to three months. In Ireland, unauthorized data access is punishable by fines and jail terms of up to three months.

Israel, Iceland, Italy, Sweden, Switzerland, the United Kingdom, and Ireland also prohibit unauthorized data access. All of those countries impose fines for such access. In addition, Israel imposes prison terms of up to three years, as does Italy. Ireland provides for imprisonment for up to three months. Sweden applies prison terms of up to two years, and the United Kingdom imparts sentences of up to six months.

Some countries impose greater penalties if the unauthorized access involves the theft of trade secrets or otherwise results in significant damage. For instance, Hungary provides for a jail term of up to three years for unauthorized data access, but increases that term to five to eight years if there are significant damages resulting from the access. Singapore imposes jail terms of up to two years; however, if significant damages are caused by the access, the prison term can reach seven years. In Denmark, prison time of up to six months is provided for unauthorized access; however, if the access involved an attempt to gain access to trade secrets, prison time of up to two years is permissible. Portugal applies a basic jail term of up to one year, but that term can be increased to five years if trade secrets are involved. Poland provides for a basic sentence of up to two years, five years if there is significant damage caused by the access, and eight years if the unauthorized access has an impact on government operations.

Criminal penalties against unauthorized access code use

A third strategy that applies criminal sanctions to reduce unauthorized computer use is to focus on the computer access systems, such as access codes, personal identification numbers, credit cards (and other financial account numbers), and passwords. Under this approach, criminal laws penalize the unauthorized distribution or use of those codes. The theory is that by controlling access to and use of the access mechanisms, unauthorized computer system use can be minimized.

The United States, Japan, and Canada all specifically prohibit unauthorized access to, use of, or distribution of passwords or other forms of access codes. In the United States penalties for abuse of access codes include fines and prison terms of up to 10 years. Japan provides for fines and prison terms of up to one year. Canada imposes fines and prison terms of up to 10 years.

Generally, these laws provide penalties for both unauthorized users and unauthorized distributors of the access mechanisms. Both the party who used the password or other mechanisms for unauthorized access and any other party who may have participated in providing that access mechanism to the unauthorized user are within the scope of most of these laws. The intermediaries would be guilty of *trafficking* in access codes, and that violation can exist even if the intermediary did not intend to assist unauthorized access. Careful management of access codes and other access mechanisms is thus very important in order to avoid liability based on inadvertent disclosure or distribution. As attention to the need for effective access code management grows, more organizations look for ways to protect those codes, including the use of on-line resources such as password security guidance available from institutions such as Cambridge University (http://www.ftp.cl.cam.ac.uk/ftp/users/rja14/tr500.pdf) and self-testing systems such as that provided by LOphtCrack (http://www.lOpht.com).

Criminal penalties for network sabotage

Several jurisdictions specifically provide for criminal penalties when a party attempts unauthorized access to a computer system for the purpose of sabotaging the computer equipment or the data it processes. In the United States, for example, federal law prohibits damage to data or computer systems that is inflicted intentionally or through reckless conduct. This means that the law prohibits both intentional sabotage or vandalism and unintentional but reckless conduct that result in damage to a computer, a network, or computer content. Penalties for that kind of action include fines and prison terms of up to 10 years.

The criminal laws of the People's Republic of China, Germany, Finland, and Italy also specifically prohibit actions that damage data or computer systems. In China, those actions are punishable by fines. In Germany, punishment includes fines and prison terms of up to five years.

In Finland, fines and jail sentences of up to one year are imposed. In Italy, fines and prison terms of up to two years are applied.

Luxembourg and the United Kingdom penalize alteration of data stored on computers. In Luxembourg, that action can receive fines or prison terms of up to two months. In the United Kingdom, the penalties include fines and prison terms of up to six months.

The laws protecting computer networks from conduct that damages the networks or the data they contain appear to be applicable to a wide range of harmful conduct. While some of those laws specifically mention that they are aimed against actions such as destruction or modification of data and introduction of viruses or worms into computer systems, they are also applicable against other forms of data system vandalism. For example, these statutes would presumably also be effective against actions such as denial-of-service attacks against Web servers.

Data theft

Access to a computer network for the purpose of stealing information is also expressly illegal in some jurisdictions. For example, Denmark imposes jail terms of up to two years for attempts to steal trade secrets. Germany considers such theft to be data espionage and punishes that conduct with fines and prison terms of up to three years. Portugal punishes theft of trade secrets with fines and jail terms of up to five years. The United States also imposes fines and imprisonment for up to 10 years.

These criminal laws may also apply to parties who use or distribute stolen data, even if they did not actually steal the material. In much the same way that a party who acquires or distributes stolen tangible goods is punished, criminal laws against data theft commonly also penalize parties involved in subsequent use or transfer of stolen data.

Civil or private law claims

In some jurisdictions, private parties can raise legal claims against each other. Generally, these legal actions involve claims that the defendant (the party against whom the case has been brought) has violated some legal right of the plaintiff (the party who raised the claim). These private claims are most common in the United States, where they include *tort* claims, as well as claims that are authorized under certain statutes enacted by legislatures (e.g., the CFAA authorizes private parties to initiate civil lawsuits when they have been harmed by violations of the act). Generally, the remedies for these civil or private law claims are court-ordered actions (e.g., injunctions) and mandatory monetary payments (e.g., damages). Court-ordered actions are generally designed to stop the illegal conduct, and the monetary awards are intended to compensate the injured party for the harm caused to it by the illegal conduct.

These private claims usually require that the party seeking relief provide evidence demonstrating several facts. If the claim is brought under rights granted by a statute, such as the CFAA, the plaintiff must show that the conduct of the defendant actually violated the terms of the statute. If the claim is a tort claim, the plaintiff must prove that the conduct of the defendant met the terms provided by law for tort liability. The plaintiff in a tort case must show that the defendant had a legal duty to perform, or refrain from performing, some act, and that the defendant, in fact, violated (breached) that duty. The plaintiff must prove that the breach of the legal duty by the defendant was the cause of harm to the plaintiff, and the plaintiff must be able to measure, or quantify, that harm. Finally, the plaintiff must be able to specify what type of relief (remedy) is necessary to compensate for the damage (e.g., payment of money or court-ordered action by the defendant).

Economic or business tort claims

Some jurisdictions, including the United States, permit recovery for damages caused by illegal interference with business activities or commercial relationships. In these cases, the plaintiff must demonstrate that the other party has acted in a manner that breaches the legal duty to avoid unfair competition and to refrain from interfering with the commercial relationships of the plaintiff. The plaintiff must also provide facts that demonstrate that the breach of the duty resulted in quantifiable harm to the business relationships or business interests of the plaintiff. Unauthorized access to computer systems or their content can sometimes form the basis for this type of legal claim.

A computer network owner or operator can raise commercial tort claims against a party who engages in unauthorized access to that network, provided that the access caused quantifiable economic harm to the network owner/operator. For example, if a hacker penetrates a company's computer network and steals valuable information, the network owner and operator would likely have strong commercial tort claims to raise against the hacker, and if those plaintiffs won a tort lawsuit against the hacker, the court would most likely order the hacker to cease the harmful conduct and pay monetary damages to the plaintiffs to compensate them for the harm caused by the hacker's conduct.

As a practical matter, parties in addition to the one directly engaged in the unauthorized access may also face legal liability under commercial tort claims. Imagine, for example, a situation in which a company provides electronic commerce transaction processing services to a retailer. If negligent conduct by the service provider permits an unauthorized party to access the transaction records of the retailer, thus compromising the security of the transaction system, the retailer will likely have sound legal support for tort claims against the service provider. The commercial tort argument would include claims that the service provider had a duty to act reasonably to secure the transactions system, and by failing to do so, the service provider's negligence resulted in commercial harm to the retailer by forcing the retailer to stop processing transactions, by undermining the relationships the retailer has with its customers, and by causing the retailer to incur some legal liability to those customers.

Legal claims against parties other than the actual perpetrator of the intrusion into the computer system are likely to become increasingly common. The parties who conduct the unauthorized access are individual people, and are therefore unlikely to be particularly wealthy. A computer system owner who has suffered significant damages as a result of the system intrusion is unlikely to be able to obtain adequate monetary compensation from such a defendant. In this setting the most likely result is that the system owner will urge government authorities to prosecute the defendant, to the extent that criminal laws are applicable to the case, and that the owner will simultaneously raise some type of civil law claim against parties such as contractors, who may have played an unintentional role in the intrusion (e.g., through negligent security services they may have provided to the network owner). This combination of pursuit of both criminal and civil sanctions will be a popular strategy, as it will facilitate more meaningful penalties such as incarceration against the actual perpetrator, and will enable the network owner to have a better chance of recovering more meaningful monetary compensation from parties who have greater assets.

Interference with property rights

Increasingly, computer equipment and the content that it processes are being looked upon as property. To the extent that the systems and their content are treated as property at law, a wide range of legal rights are enforceable by the owners of that property under traditional property law principles. Two examples of private claims that can be raised in response to violations of property rights are theft and trespass (described as "conversion" in some jurisdictions). Theft consists of taking unauthorized possession of property owned by another party. In many jurisdictions, theft is both a criminal law violation and a civil law claim. Trespass involves use of property owned by another party without the permission of that owner.

Computer system operators now treat both their equipment and the content of their systems as their property. Based on these property law theories, unauthorized use of equipment or data content is the basis for a private lawsuit seeking monetary compensation. In the digital environment, some now argue that property includes electronic material, including computer files and the information they contain. Under this approach, an owner of a Web site, database, or other computer content would treat that content as his or her personal property and would seek compensation from any party who used that property in a manner inconsistent with authorizations granted by the property owner.

Interesting examples of this argument have emerged in the context of Web-search systems and directories. In the United States, for instance, there are several lawsuits that raise these property law arguments (in addition to other legal claims), including disputes between eBay and Bidder's Edge, in the context of searches involving on-line auctions, and between the House of Blues and Streambox, in the context of access to streaming media content.

In the *eBay/Bidder's Edge* case, the property at issue is the data files that contain information about on-line auctions processed through the eBay system. The alleged interference with that property is the use of "shopbots" by Bidder's Edge to enable buyers to access information about products available through eBay auctions as part of the Bidder's Edge search capability. eBay contends that the data files and the auction information they contain are property owned by eBay and accessible only to the parties invited to use that information by eBay.

The basic property law question is whether data files, in this context, qualify for protection under traditional personal property legal principles and whether the owner of such property can assert complete control over access to the property. At present, this case remains unresolved, but the property law issues it raises have significant potential impact on the on-line search function, in general. If on-line information and data files are property subject to the complete control of their owners, those owners can restrict access to that material in any way they see fit. That approach could severely restrict the ability of Internet users to gain access to the diverse content they seek.

The *House of Blues* dispute with Streambox focuses on access to streaming media content. The Streambox on-line search and directory capability enables people to find streaming media content that is of interest to them quickly and easily. After finding the content, the user is linked directly to the streaming content he or she wanted. Included among that content was streaming audio and video of musical concerts that had taken place at House of Blues' restaurants and was accessible through the House of Blues' Web site.

The House of Blues has gone to court to try to stop Streambox from including its content in the Streambox search and directory functions. One of the legal principles cited by the House of Blues in support of its claim is the assertion that the data files that contain the media content are House of Blues' property and that access to and use of that property by Streambox violates the limits on use that House of Blues has established for its property. This case also raises the issues regarding whether traditional property law rights are applicable to data system content and the extent to which owners of that content can control access to and use of that property.

Recognize that these property law arguments are not traditional intellectual property law claims (e.g., copyright infringement). The new property law arguments may be used to supplement traditional intellectual property law claims, but they are distinct legal arguments. Intellectual property law claims rely on the specialized set of laws that has been developed to protect and to manage copyrights, patents, trademarks, and trade secrets. The new property law theories attempt to apply the well-established principles of law that govern the transfer and use of personal property to computer equipment and the electronic content of that equipment.

These cases also raise the issue of management of access to computer systems by intelligent agents and other surrogates or proxies for humans. As the capabilities of those agents increase, their ability to gain access to computer systems and to take actions affecting those systems (and their content) also becomes more sophisticated. Human beings continue to be held legally accountable for the conduct of their electronic surrogates. If my intelligent agent engages in unauthorized access to a computer network, I am responsible for that violation. As our electronic proxies become more active and more independent, we would do well to remember that it is the human operator of the intelligent agents who will be held accountable for any violations of law by those agents.

The controversy associated with these cases also suggests to some observers that self-help measures to manage access to on-line content may be preferable to a rush to litigation. There may be technical measures that an owner of digital property can take to minimize the risk of unwanted access to that property by other parties. For example, systems that permit access to Web content only after registration can help to keep automated search programs from obtaining that content. If content providers could rely more on the technical and business measures they can take on their own, perhaps there would be less need for court and other legal action. The advantage to the self-help approach is that it permits owners of the material to manage access as they see fit, but it does not set formal precedents that will govern the future actions of other parties, as does formal court or regulatory action.

Controlling access to protect trade secrets

Ineffective network access controls can undermine the ability of an organization to protect its trade secrets. Weak access management makes it more likely that trade secrets will be stolen or otherwise compromised. In addition, poor computer system access management can result in the loss of the ability of an organization to enforce trade secrets protection in the future.

The law of trade secrets permits businesses to protect information that has competitive commercial value from public disclosure. It requires, however, that a party claiming trade secrets protection must do all that it reasonably can to guard those secrets from disclosure. If an owner of trade secrets has acted diligently to keep that material protected from public disclosure, but the secrets are nonetheless disclosed, through no fault of their owner, that owner has the legal right to ask a court to stop the disclosure and to compensate the owner for harm caused by the unauthorized disclosure. In contrast, if the owner of the secrets did not take precautions to preserve the secrecy of the material, courts will not protect the material as trade secrets; thus if they are disclosed, the court will not stop the disclosure and will not award compensation to their owner.

If your computer system contains material that you want to protect as trade secrets, you must implement effective access controls. If the secrets are ever compromised, those access controls will be necessary to demonstrate to a court that your organization did, in fact, treat the material as secret and that you are, accordingly, entitled to legal remedies as a result of their disclosure. Computer system access controls thus play a major role in preserving your organization's legal right to protect its trade secrets.

The challenge of managing computer network access in a manner consistent with preservation of trade secret legal rights becomes more difficult as organizations want to share more of their network content and services with other parties. As companies interconnect their networks with their business and trading partners (e.g., retailers linking their computer systems with those of their suppliers and product manufacturers) and with their customers (e.g., product manufacturers linking their networks with the distributors of their products), it becomes more difficult to control system access. In this environment of many different authorized users, interconnected networks owned by different parties, and different types of information accessible to different groups of system users, extreme caution in managing the operations of the system is essential to preserve trade secrecy protection.

If your organization fails to provide computer security adequate to protect the trade secrets of other parties that you have been authorized to store or use, you and your organization may face legal liability. The owner of the trade secrets could, for instance, raise breach of contract law claims against you, arguing that your failure to protect the trade secrets was a violation of your contract obligations under a confidentiality or nondisclosure agreement. The trade secrets owner could also raise a commercial tort claim against you for misuse of the trade secrets or a claim of unfair business practice resulting in the disclosure of the secrets. Therefore, you must treat the trade secrets of other parties with the same care regarding security that you apply to your own trade secrets.

Unsolicited commercial e-mail: The spam problem

One common form of unauthorized computer system access that has generated a substantial amount of legal attention is the issue of unsolicited commercial e-mail (*spam*). Many ISPs have implemented procedures to restrict access to their networks by parties distributing spam. Various jurisdictions have enacted laws that restrict or prohibit the distribution of these unsolicited commercial messages. In effect, when a commercial e-mailer sends multiple unsolicited messages into an ISP's network after the ISP or legal authorities have prohibited such mailings, the act is one of unauthorized network access. The struggle to limit spam is an example of the complex nature of computer system access restrictions.

ISPs seek to limit spam for at least two reasons. The substantial volume of the commercial messages places great operational demands on their networks. That demand sometimes results in additional operating expense and degraded service quality. Spam is also irritating to many of the customers of the ISPs, and in an effort to please those customers, many ISPs try to reduce or eliminate those unwanted messages.

ISPs have various legal claims at their disposal for use against spam distributors. Laws enacted by several states in the United States (e.g., California, Washington, and Virginia), give ISPs the right to raise civil law claims against those distributors. Those laws give courts the authority to order the spam distributors to stop sending the unsolicited messages through the ISP networks and the authority to order the distributors to pay monetary damages to compensate the ISPs for damage to their network operations.

In jurisdictions where there are no specific statutes regulating distribution of unsolicited commercial e-mail, other legal claims have been raised by ISPs against the spam distributors. For instance, commercial tort claims alleging that the commercial messages harm the ISP networks and harass ISP customers, thus resulting in adverse economic impact on the ISPs and damage to the business relationships between the ISPs and their customers, are popular legal claims. Some ISPs may also raise property law claims, asserting that the actions of the e-mail distributors damage the servers and other property of the ISPs (by overloading the networks and degrading the quality of service), and seek compensation for that damage. ISPs also raise breach of contract claims, alleging that unsolicited e-mail distribution through their networks violates the terms of service under which the ISPs carry those messages, and that violation constitutes a breach of contract.

Some of the antispam laws enacted in the United States have been challenged as violations of legal rights granted by the U.S. Constitution. Several different constitutional law claims have been raised. One claim is that the restrictions on these commercial messages violate rights of free expression provided by the First Amendment of the U.S. Constitution. The Constitution does provide protection against laws that unreasonably impair the ability of businesses to distribute "commercial speech." The unresolved question is the extent to which those laws make antispam statutes unconstitutional.

A second claim is that some of the state laws enacted against spam result in unreasonable state-imposed burdens on "interstate commerce" in the United States, and are thus violations of the Commerce Clause of the U.S. Constitution. This argument is applied to state laws that bar unsolicited commercial e-mail that is generated in another state. Some courts have concluded that a ban against commercial e-mail created in another state is unlawful as it creates an unreasonable impediment to interstate commerce.

These constitutional law challenges in the United States have not yet been fully resolved. For now, the state laws against spam continue to be in effect and more are under consideration. The constitutional challenges directed toward some of those laws must, however, be resolved before we can define clearly the extent to which legislation banning or restricting spam provides a long-term solution.

Spam distributors who have had their traffic blocked by ISPs have raised their own legal arguments to challenge that denial of service. In some instances, the commercial e-mailers argued that the prohibitions against their traffic were anticompetitive actions that constituted unfair trade practices in violation of antitrust and competition laws. These arguments have not, to date, been successful. While there are certainly circumstances under which denial of access by ISPs could be illegal, as violations of antitrust laws or competition law requirements, commercial e-mailers have not been particularly successful with that argument, to date.

Another potential claim available to commercial e-mail distributors is that their messages do not qualify as spam. For example, an e-mailer that distributes commercial messages only to parties who have invited such communications may be able to be successful in a legal challenge to an ISP that blocks its messages. ISPs should exercise caution to make sure that their efforts to reduce unsolicited commercial messages do not prevent distribution of other communications. In that type of situation, an e-mailer may be able to raise commercial tort claims for damages suffered as a result of the message blocking or it may be able to raise a defamation claim, arguing that the ISP improperly labeled the e-mailer as a spam-sender, and that characterization harmed the e-mailer's commercial reputation.

E-mail distributors have also challenged parties who try to assist ISPs to limit the volume of commercial e-mail. For example, in the United States, the Mail Abuse Prevention System (MAPS) has been sued by different e-mail distributors. MAPS maintains a list of commercial e-mail distributors, and various ISPs use that list as a basis for their traffic-blocking decisions. Some ISPs reportedly block traffic from senders identified on the MAPS list. Several parties have sued MAPS, arguing that they should not have been included on the MAPS list, and that their e-mail traffic should not have been blocked. Those cases illustrate how intermediaries involved in spam-blocking decisions may also face legal claims from the commercial e-mail distributors.

The "zombie" network problem

Highly publicized distributed denial-of-service attacks that made use of computer equipment and networks owned by third parties raise a legal issue associated with potential liability when computer equipment is used without the equipment owner's authorization in order to cause damage to computer systems or data owned by other parties. When a computer system is functionally commandeered (i.e., made a *zombie* network) and is used as an instrument to attack another system, challenging issues of potential legal liability emerge.

The owner of the zombie network can make use of business tort claims to seek compensation for damage caused to its network or business. The argument would likely consist of a claim that the intentional conduct of the party initiating the attack resulted in harm to the zombie network and damage to the commercial interests of the owner of that network. The owner of the commandeered network may also be able to raise claims under computer access laws that provide for civil or private remedies for unauthorized access. The owners may also be able to raise property law claims based on the theory that the network is their property and the unauthorized user made use of that property without their permission.

In jurisdictions in which there are criminal laws prohibiting unauthorized access to a computer system, law-enforcement authorities could prosecute the unauthorized user. In those actions, the government would take the claim to court and the party who commandeered the network would be the criminal defendant. In some jurisdictions, prosecution requires a demonstration that the network that was compromised was protected by some form of access control or security measure which the defendant circumvented. In those jurisdictions, successful prosecution of the intruder would require that the operator of the zombie network had the requisite security measures in place prior to the attack.

It is also possible that the owner of the zombie network may be liable for damages under a private claim raised by parties who were harmed by the attack. For example, if an unauthorized party commandeers a computer system owned by another party, and uses that system as a tool to facilitate a denial-of-service attack against a commercial Web site, the owner of the commercial site that is the target of the attack may have a basis for a civil lawsuit seeking compensation from the owner of the system that was commandeered for harm to the commercial Web site and its business operations.

The best way to reduce the prospects of this type of claim against the owner of the zombie network is through use of effective network management and security systems and practices. If the owner of the zombie network can demonstrate that it applied a reasonable level of security to protect its computer system from unauthorized access, the owner will have a stronger case to raise against liability claims from parties injured by the attack. In contrast, if the owner of the zombie network did not apply generally accepted security measures to protect its network, there is a better chance that injured parties may be able to win claims against the network owner.

A court would likely evaluate the security practices of the network owner in order to assess whether that owner acted reasonably in its efforts to secure its network from the type of misuse that resulted in the harm to the injured parties. As computer security and network management technologies improve, computer system owners will be expected to apply those advances to their systems. While not always obligated to use the most modern security measures, courts will increasingly expect system owners to make prudent use of the most effective available security measures to guard against the most likely threats to the system. Denial-of-service attacks using zombie networks are now widely recognized as an important computer security threat, and in that environment, courts will expect computer system operators to secure their networks against that threat.

Increasing use of contractors or other outside parties to manage and maintain computer systems, in some form of outsourced service arrangement, raises another interesting liability possibility. If the computer system that was commandeered was being managed or operated by a contractor, it is possible that the contractor may face liability. The owner of the system (i.e., the contractor's client) could raise a claim of breach of contract against the contractor. Parties harmed by the denial-of-service attack could sue both the owner of the zombie network and the contractor tasked with operating the network for damages based on a commercial tort law claim.

With these potential legal liabilities in mind, some computer network operators, including some ISPs, are now working together to try to share information and develop the best practices to help keep their networks from becoming unwitting accomplices in denial-of-service attacks. An example of this effort is the RFC 2267-plus Working Group, a consortium created to try to reduce the frequency of these attacks and to help legitimate networks minimize the risk that they may face some type of liability if they are unintentional participants in such an attack.

Operators of large computer networks should pay particular attention to the risk of this type of liability, and among those large networks, perhaps the most exposed to the zombie problem are university computer networks that often operate without the same level of security as large corporate networks. Also recognize, however, that smaller networks must now face this same issue, as more and more of those networks increase their exposure to unauthorized access through use of higher capacity, full-time Internet connections.

Access in an environment of outsourcing IT functions

As many organizations now retain other parties to manage and operate all or part of their computer system functions, access control issues are now more complex. The party who owns a computer system and its content may no longer be the same party that controls access to the system. In some outsource relationships, control over access to a computer system is a function that has been delegated by the owner of the system to another party. This delegation raises some legal liability issues. At one level, it introduces another party who may bear some liability for unauthorized system access. For example, if a customer's credit card information is stolen from the computer system of a merchant, the customer could raise private legal claims against the party who stole the information, the party who owned the computer system, and the operator of that system. In many instances, the owner and the operator of the computer system are the same party, but when computer system operations have been moved to the control of a contractor, that contractor and the system owner can both face legal liability for negligent conduct that may have facilitated the unauthorized access.

Outsourcing permits the ownership and operational management functions of computer systems to be handled by two different parties. Under those circumstances, each of those parties may bear legal liability if their conduct contributes to unauthorized system access that harms another party. The existence of a contractor does not relieve the owner of the computer network from legal liability. As noted above, in most instances, an injured third party would have legal claims against both the contractor and the owner of the network.

Contractors handling computer services for their clients may also face legal liability to those clients under contract law principles. For example, a contractor who provides network security services for its clients may be sued for breach of contract if unauthorized access to the client's computer system results from a failure of the contractor to perform all of the duties specified by the contract between those parties. In such actions, the client (the computer system owner) is the plaintiff and the contractor is the defendant.

Contractors may also face tort law claims from their clients in the event that there is damage caused by unauthorized network access. For instance, imagine a contractor retained to provide, develop, and manage a commercial database for his or her client. Assume that unauthorized access to that database results in the theft of information that constitutes trade secrets of the client. The contractor could be sued by the client under a business tort claim, with the client essentially arguing that the negligence of the contractor in performing his or her services resulted in loss of the trade secrets and corresponding economic harm to the client's business. This argument would assert that, while the contractor did not steal the trade secrets, his or her conduct was inadequate, breaching a basic duty of care owed by the contractor to the client and contributing to the theft.

In this environment, it is particularly important that information technology outsourced service providers and their clients reach, in advance, a clear understanding of their relative obligations regarding computer system access security. Those obligations include duties and performance commitments, as well as agreements regarding the sharing of legal liability in the event of unauthorized access. The legal liability is, at least in part, an issue of risk-sharing, a determination of how much of the potential cost associated with network security breaches will be absorbed by the client and how much will be absorbed by the provider of the outsourced services. A written contract that clearly identifies and defines these obligations related to security (e.g., a service-level agreement) is an essential element of effective security management.

Unauthorized access in the pervasive computing environment

As the variety of computing devices increases and they become ever more widely dispersed within other products (e.g., electronics equipment, appliances, vehicles, clothing), the number of legal claims based on unauthorized computer access will almost certainly increase. To date, we have seen criminal and civil laws pertaining to unauthorized computer use applied almost exclusively in the context of traditional computer equipment (e.g., servers, PCs). In the future, those same laws will likely be applied to an ever-increasing number of different types of computers, many of which will be components of noncomputer products (e.g., consumer appliances).

For example, assume your automobile has an onboard tracking system that uses intelligent microprocessors and other computational devices to monitor and record your travels. Then imagine that, while your automobile is being serviced, one of the mechanics accesses that stored information without your permission. In many jurisdictions that have criminal or civil laws prohibiting unauthorized computer access, that action would be a violation of those laws.

As computers and computer networks become more pervasive, it follows that the laws controlling access to and use of computing devices will be applied far more frequently than they are today. The number of contexts in which those laws will be applied will also increase, leading us to an environment in which computer access and use laws may be called into play in a large number of cases which would not appear, at first glance, to be computer crimes at all. It may not be long before we begin to see cases, for instance, in which a teenage vandal who damages a pump at a gasoline station may be prosecuted for a federal computer crime simply because the teenager had the misfortune of selecting a "smart" pump as his target. In that kind of setting, many individuals and businesses not directly involved in traditional information technology markets or industries must nevertheless be familiar with the legal rights and liabilities associated with computer access and use laws.

Access by government authorities

Increasing attention is now focused on the issue of access to computer and communications networks by government authorities, particularly law enforcement and national security agencies. At issue here is the question of which circumstances justify permitting those authorities to monitor computer and telecommunications systems through use of monitoring devices. Most people acknowledge that there are some circumstances under which it is in the public interest to permit certain government authorities to obtain access to private computer and communications network content. Most people also recognize, however, that such government access should be carefully constrained to avoid interference with basic legal principles of individual freedom and personal privacy. Laws in most jurisdictions permit government monitoring of computer networks, under some circumstances. The limitations applied to that network access are, however, widely varied.

In the United States, for example, controversy arose with respect to the Federal Bureau of Investigation's (FBI) use of a surveillance system to monitor Internet traffic. Known originally as Carnivore (now sometimes referred to as the DSC 1000 system, a name that presumably is intended to make the system seem less threatening), the technology enabled authorities to search Internet usage (including e-mail and Web browsing) of targeted individuals. An important issue associated with computer network surveillance systems such as Carnivore is the extent to which computer network operators will be obligated, by law, to cooperate with the authorities when those authorities want to conduct surveillance. Should the law obligate computer network owners to make their systems technically compatible with the government surveillance systems so that the authorities can conduct surveillance when they have the requisite legal authority to do so? Who should bear the costs associated with any network modifications necessary to make the network compatible with the surveillance technology? Should ISPs be required to allow government authorities to place surveillance technology in the ISP networks? At present, those policy issues remain unresolved in many jurisdictions.

Another example of controversy over computer network monitoring by government authorities is provided by the RIP legislation in the United Kingdom. One of the requirements of RIP is that all Internet traffic passing through the United Kingdom be routed through server facilities controlled by the national security agency of the U.K. government (MI5). Some parties have objected to this requirement, asserting that this requirement provides a platform to facilitate content monitoring by the U.K. authorities.

Both surveillance systems noted above require the issuance of valid legal orders (e.g., *search warrants*) prior to use. Yet various parties remain concerned that the two government initiatives represent unreasonable government intrusion into computer system content and operations. Critics argue, in effect, that Carnivore and RIP are examples of excessive government access into computer networks. The fear is that such access can too easily be abused by governments, harming the civil liberties and individual rights granted by law to the citizens of the United States and the United Kingdom. It is worth noting that both the United States and the United Kingdom have long histories of respect for civil liberties, and to the extent that there is concern about potential abuse of computer system monitoring authority by governments in those nations, imagine the potential threat to those liberties posed by similar digital surveillance initiatives by governments in other countries that do not have the same tradition.

Other critics suggest that systems such as Carnivore, that make use of surveillance technology integrated into the public data networks, undermine network security in ways other than the government surveillance they support. Some claim that the presence of *black boxes* that collect and analyze data on network content and use will provide an irresistible target for private parties who would like to have access to that information but who are not authorized to obtain it. In effect, these critics contend that introduction of monitoring systems into the public network may simply provide greater incentive for unauthorized users to try to break through the security of the network in order to gain access to the attractive information collected by the government surveillance systems.

Computer system surveillance conducted by governments also raises important potential conflicts among governments. For instance, there is an ongoing international dispute associated with the Project Echelon system reportedly operated by the intelligence agencies of the U.S. government. Critics of the system allege that it is a global interception and relay system that collects communications content in bulk, and searches that content for information of interest to the intelligence agencies. Some allege that the U.S. intelligence agencies that operate the system share the content and search results with intelligence agencies in other countries (the United Kingdom, Canada, Australia, and New Zealand, for example).

Critics believe that the communications monitored by the Echelon system include Internet communications (e-mail messages, for instance). Concern about the system has been expressed by legislatures in countries such as Denmark and Italy. Private companies have also expressed concern, fearing that the communications that are monitored contain commercially sensitive information, which, if disclosed to U.S. companies, could give those companies a competitive advantage over businesses in other countries. Governments and companies outside of the United States are concerned that the system could be used to conduct economic espionage.

European governments are also moving to preserve their access to telecommunications system (including Internet) content. The Council of Europe is considering mandatory content access rules that are advocated by law-enforcement authorities in Europe (Council of Europe Conclusions, ENFOPOL 23, 30.3.01). The proposed European policy would require that telecommunications information be retained for a period of at least seven years, and that all such stored information be accessible to law-enforcement authorities. There is apparently still significant uncertainty as to what telecommunications information would be within the scope of the requirement (although it seems that e-mail messages and Web content would be included) and which legal standards, if any, law-enforcement authorities would be required to meet before obtaining access to that stored information. Would a valid legal order, a court order for example, be required, or would a simple demand by the law-enforcement authority be sufficient?

Enactment of this type of regulation would have a significant impact on telecommunications service providers as well as computer system operators and users. In addition to the obvious legal issues associated with disclosure of this information to government authorities, mandatory retention of this material also raises serious potential legal liability to private parties. For example, to the extent that some of these records constitute private personal information or communications message content (e.g., e-mail messages), there are already specific legal requirements associated with security for, and disclosure of, that material (e.g., the EC's Information Privacy Directive).

By requiring retention of that material, governments are essentially forcing private parties to increase the amount of content that they have strict legal obligations to secure. This will almost certainly increase the cost of legal compliance for those parties, and will likely increase the chances that they will face future liability for security breaches. Mandatory retention of this information will also likely provide greater incentive for parties involved in private legal actions to use the evidence discovery process associated with litigation to gain access to this sensitive material. Litigants will know that these records of communications content exist and they are likely to view them as an extremely attractive source of evidence during the litigation process, as well as a valuable source of competitive commercial information accessible through legal discovery.

Mandatory access

The discussion above dealt with one form of mandatory access, network access by government authorities under legitimate legal authorization. There are also some emerging examples of instances in which the law requires access to a computer system for certain private parties, as well. There is increasing concern among some parties about laws that require certain private parties to be given access to the computer systems of other private parties, under limited circumstances. One example now being addressed in the United States arises in the context of the Uniform Computer Information Transactions Act (UCITA), a model statute that has been enacted by some states (e.g., Maryland and Virginia) in the

United States and is currently under consideration in several others. UCITA is intended to provide a common set of legal standards to be applied in the various states to govern the rights and obligations of buyers and sellers of computer software and other information products.

Part of the debate over UCITA focuses on a provision in the model statute that permits sellers of software to build mechanisms into their products that would enable remote disabling of the software. The notion is that the seller would have the right to use that capability to disable the software if the buyer did not meet payment or use obligations imposed by the seller as conditions of the sale. In effect, this right granted to a seller of software products would provide a legally enforceable right of access to the software, even after it has been purchased and installed by the buyer. That right effectively constitutes mandatory access to the buyer's network for purposes of disabling the software, even though that right could be exercised from a distance, outside of the network.

Not surprisingly, software developers, on the whole, support the UCITA provision, while software users generally object to it. Each state is free to adopt or reject UCITA, and even those that adopt the basic UCITA form would be free to modify any of its provisions, prior to enactment. It is currently unclear how many states will ultimately accept this provision, but UCITA nevertheless provides an example of a law that specifically permits at least a limited form of mandatory access to a private party's computer system.

Another example of this type of mandatory network access is the *open access* obligation that is developing in the United States. This requirement has been imposed on certain cable system operators and requires those operators to permit all ISPs to have access to the customers of the cable systems. This concept was an important aspect, for example, of the regulatory approval process for the merger of America Online (AOL) with Time Warner. Open access requirements are intended to ensure that Internet users have choices in regard to the service providers they select. The requirement is intended as a policy measure to ensure that users have diverse methods to access the Internet, but it also involves a form of mandatory access to the networks of the cable service providers. In order to make sure that end users have access to multiple ISPs, the U.S. government has been willing to require a form of mandatory access by ISPs to the networks of the cable service providers.

Another example of a mandatory access obligation is being explored in the context of Internet users who have some form of physical disability. The argument raised by some advocates for the physically challenged is that the Americans with Disabilities Act (ADA) legislation in effect in the United States requires ISPs to provide reasonable accommodations to enable blind people to access Internet content. An example of this concept was provided in a lawsuit filed by the American Federation for the Blind against AOL. In that suit, the federation argued that the ADA required that reasonable accommodations be made to make public places accessible to people with physical disabilities.

That case reached at least a temporary settlement, in which AOL agreed to make use of software that will facilitate Internet access by people who are blind; thus the court has not yet been called upon to make a formal legal interpretation. The case illustrates, however, a legal argument that will likely be raised against other ISPs. Based on future court interpretations, the actions that AOL has voluntarily agreed to make in this context could become mandatory at some point in the future. Should that happen, the law would effectively impose an access obligation on behalf of blind Internet users.

For the future, some access obligations may arise as a result of the increasing concern about the so-called "digital divide." Some policymakers are deeply concerned that there are large segments of the population in all parts of the world that do not have effective access to Internet content. Concern that those who are economically poor or who live in isolated regions may be denied the benefits of Internet access has been widely expressed. We are likely to see policy initiatives in many different parts of the world that will be aimed toward *bridging* the digital divide by assisting those who have so far been denied Internet access to obtain such access. At least some of those policy initiatives may involve some form of mandatory access obligation on the part of ISPs.

For its public telephone network, the United States has long applied a public policy of promoting *universal service*. That policy aims to make basic telephone service available to all U.S. citizens, no matter where they live. Some leaders advocate adoption of a similar policy objective for Internet access. If a form of universal service obligation is ever developed for Internet access, it would place some service requirements on the ISPs. Those requirements would mandate service to certain areas and certain groups of users. Such an approach would constitute another form of legally required computer network access. They may also carry with them obligations for payments, in the form of fees or taxes to be used to pay for the network and service expansion necessary to support the expanded service offerings.

Appendix 3A: Computer system access guidelines

All computer system owners and operators should establish and enforce access guidelines for users of their systems. The guidelines should be put in written form and should be provided to all authorized users of the computer system. In addition to the written documents, training sessions should be provided for all system users to make them familiar with the guidelines. Violations of the access guidelines should result in denial-of-system-access privileges, termination of employment or prosecution under civil or criminal laws. The existence of guidelines can help to reduce the chances of an access problem and can help limit any legal liability associated with an access breach. At a minimum, those guidelines should include the following topics.

Designate an employee to be responsible for system security

The organization should identify a person or position which will be responsible for management of computer system access security. Some organizations have a chief security officer whose duties may include these functions. Others may have their chief information officer or chief technology officer perform this role.

Identify authorized users

Guidelines should clearly identify those individuals who are authorized to access the computer equipment and systems operated by the organization. The guidelines should also identify limits to authorized use. Access authorizations should be granted only to those parties for whom there are legitimate business reasons supporting such grants. Part of the process of identifying authorized users should include thorough background checks on all parties who will have access to sensitive information.

Protect the physical security of equipment

The guidelines should specify measures to be applied to protect the physical security of computer equipment and to control access to that equipment. The physical security measures should include equipment monitoring/inventory, secure sites for computer equipment, and requirements that equipment be shut down when not in use.

Access controls

Access controls (e.g., codes, passwords, biometric controls) should be used for the system, and all system users should be required to make use of those controls. The password or other access control for each authorized user should be unique to that user. Use of default passwords provided by equipment suppliers should be prohibited. System users should be required, for example, to surrender their access codes and passwords upon order from management. The guidelines should indicate that sharing or other distribution of access information is unacceptable conduct.

Encryption

Encryption systems should be made a standard part of the network, protecting all communications and content.

Firewalls

Firewalls for network protection should be installed, maintained, and updated regularly.

Virus protection

Antivirus protection should be installed and kept current.

Access monitoring

Install and use systems that permit monitoring of system access and use. Several commercial software products are available to help manage network monitoring functions. Examples of such products include: EnCase (http://www.encase.com), SmartWatch (http://www.wetstone-tech.com), SilentRunner (http://www.raytheon.com), Forensic Tool Kit (http://www.accessdata.com), and Net Threat Analyzer (http://www.forensics-intl.com).

Intrusion response

Develop an intrusion response plan to be followed immediately when unauthorized access has been detected.

System testing

Access security systems should be tested on a regular basis. If outside parties are used for such testing, those parties should be subject to a thorough security screening in advance of performance of the work.

Appendix 3B: Sharing liability between contractors and clients

When all or part of the management of computer system access has been transferred to another party (e.g., a contractor), both the owner of the system and the contractor face legal liability for unauthorized access events. It is in the best interest of both of those parties to implement methods to share that liability. Effective methods to accomplish that goal include those outlined below.

Create a written contract

Make sure that the terms of your understanding with the contractor are clearly expressed in a written contract (e.g., a service-level agreement).

Define responsibilities

The agreement should define the security responsibilities of both the contractor and the client. The definition of responsibilities should include a clear description of performance obligations, including performance levels.

Establish penalties

The parties should agree on which penalties (e.g., payments or price discounts) will be applied in the event of failure to meet responsibilities established by the contract.

Sharing risk

The agreement should address the issue of how the two parties will share costs incurred by the parties as a result of any security breach.

Third party liability

Although a difficult issue, the parties should try to agree in advance as to how legal claims raised by third parties as a result of a system security breach will be handled. There is no way to prevent an injured third party from bringing legal action against both the client and the contractor, but both parties can try to plan how they will coordinate their efforts in response to any such claims. That coordination should include agreement regarding how the parties will coordinate their defense strategies in the event of litigation (e.g., how they will share information) and how they will each work with any law-enforcement authorities who may be involved in an investigation resulting from a security breach. In some instances, it may be appropriate for one party to agree to *indemnify* the other if that other party is required to incur costs as a result of liability. An indemnity is basically a promise that, if under certain predefined circumstances the indemnified party is forced to bear some cost (e.g., pay monetary damages ordered by a court), then the indemnifying party (i.e., the one making the promise) will reimburse the indemnified party for those costs.

Appendix 3C: Minimizing potential liability associated with commercial e-mail

Widespread distribution of unsolicited commercial e-mail can result in legal liability. Some jurisdictions either prohibit unsolicited commercial e-mail or place restrictions on such messages. Senders who do not comply with those legal requirements can be subject to fines, civil damage awards, or court injunctions. The actions described below will help to reduce the risk of that liability.

Obtain permission/consent

The safest approach to commercial e-mail distribution is to obtain permission from the intended recipient prior to sending the message. If prior permission (consent) is obtained, legal liability for the sender can be avoided.

Avoid misleading conduct

Commercial e-mail messages should clearly indicate that they are commercial communications and they should accurately identify the sender. Messages that disguise the sending party or mislead the recipient regarding content are illegal in some jurisdictions and are unacceptable.

Provide for feedback from the recipient

Commercial messages should provide a clear process through which the recipient can respond directly to the sender. That process should clearly explain how the recipient can have his or her name removed from the mailing list.

Abide by ISP restrictions

Parties sending commercial e-mail messages should be aware of the policies on such messages established by all ISPs, and they should comply with those policies. Failure to do so can make the sender liable for claims raised by the ISPs under antispam statutes or under other civil law principles.

Appendix 3D: Managing unauthorized access events

In spite of careful planning and diligent enforcement of practices and procedures, it is likely that attempts at unauthorized access of your computer system will occur. It is important to implement procedures when unauthorized access is detected, in order to minimize potential legal liability and to preserve as many of your organization's legal rights as possible. Those procedures should include the ones listed below.

Terminate the intrusion

Organizations should have a specific response prepared so that active system intrusion can be terminated. A response plan should be directed toward termination of the unauthorized access.

Minimize system damage

A plan should be in place to respond to unauthorized access in ways that minimize the amount of damage to the network and its content caused by the unauthorized access.

Assess damage

The response plan should also provide a mechanism through which the amount of damage caused to the computer network and its content can be evaluated promptly. This assessment can be used as the basis for future legal claims.

Document the incident

Systems and procedures should be in place to enable the organization to document the intrusion, the response to that intrusion, and the damages incurred as a result of the intrusion. The record of the incident and the organization's response to the incident may be an important part of evidence in future legal proceedings.

Protect the incident documentation

All documentation associated with the incident should be treated as the organization's important records. Those records should be maintained securely so that an effective chain-of-custody can be established, enhancing their value as legal evidence.

Reporting

An effective response plan should clearly describe how and to whom information regarding the incident will be reported. The plan should deal with incident reports to employees of the organization, investors, business partners, and law-enforcement authorities. Plan in advance whom you will inform and what type of information you will disclose to each party.

Facilitate the investigation

A response plan should provide for effective investigation of the event. This may involve cooperation with law-enforcement authorities or performance of some form of investigation by the organization itself. Several private organizations now provide *forensic* services to support investigations of computer security problems. Examples of such organizations are: New Technologies, KPMG, Riptech, and EDS.

Appendix 3E: Insurance as a means of limiting legal risk

Insurance coverage to guard against unauthorized computer system access and use is now available. Companies such as Insuretrust.com (http://www.insuretrust.com), Ace Ltd. (http://www.acelimited.com), and J.S. Wurzler Underwriting Manager's, Inc. (http://www.jswum.com) are examples of insurers providing coverage in this field. For many system operators, it provides an effective addition to an overall computer security strategy. The following topics are significant to both providers and buyers of this type of coverage.

Scope of coverage

As with any form of insurance, the scope of coverage is the most essential aspect of computer system insurance. Buyer and seller must agree on the events that are covered by the policy. An important first step is to check your organization's existing conventional insurance coverage (e.g., coverage for property, business interruption, errors and omissions). In some instances, that current coverage may be applicable to computer equipment and content. At least one court in the United States has concluded that an insurance policy that covers computer equipment may also cover at least some of the data stored on that equipment. You should first conclude what coverage, if any, your traditional business insurance policies may provide to your computer systems.

If the traditional insurance coverage does not provide adequate protection, you may want to consider insurance coverage that is specifically directed toward computers and their content. Important issues regarding the scope for such coverage include: Does the policy cover damages caused by both deliberate and unintentional security breaches? Does it cover damage caused by both unauthorized and authorized users? Does it cover both direct and indirect damages resulting from unauthorized access? How does the policy deal with multiple incidents? What preventative measures are required?

Premiums

Obviously one of the key aspects of insurance in this area is the price of that coverage. Premiums may vary significantly based on the scope of coverage and the risk profile of the insured party.

Term and payout value

The length of the policy and the dollar limits to payouts are of major importance. Also, the timing and process for payouts should be considered.

Security standards

Most of these policies require some form of operational standards to be used by the operator of the insured network. Such operational standards commonly include: rigorous access controls (e.g., access codes, passwords), clear definitions of system use authority/access procedures, full-time access monitoring systems, and effective compliance and enforcement policies and practices.

Security testing

Insurers commonly perform periodic inspections of the insured party's security systems and procedures. The frequency, form, and scope of those inspection tests are important elements of the coverage. Some organizations make use of third-party certification to obtain better rates from insurers and to increase the confidence of their business partners and investors. These third parties test security systems and practices and provide independent assessments of the security level for the computer systems they review. Organizations providing such security reviews and certifications include: TRUST-e (http://www.truste.com), Ernst & Young (http://www.ey.com), BBBOnline (http://www.bbbonline.org), and Tru-Secure (http://www.trusecure.com).

Appendix 3F: Legal aspects of access security testing and disclosures

Efforts to develop effective access security require testing of security measures and sharing of information with other parties. Some of the more aggressive testing methods include deliberate use of computer hackers to probe systems. In regard to information sharing, there is currently a reluctance to disclose information regarding security breaches. Parties considering use of hackers for testing and those reluctant to share security information should consider some of the potential legal aspects of those actions.

Hackers as security consultants

Some organizations now turn to "reformed" hackers to test their computer security systems. These "ethical" hackers are often retained by organizations to probe their systems for weaknesses, performing network vulnerability assessments. The services provided by these consultants are often effective and valuable. Their clients should also be aware, however, that there may be instances in which use of consultants with this type of background could cause future legal problems. If, for example, a provider of content stored on the network ever raises a claim against the network owner for misuse of that content, even if the hacker consultant did nothing wrong, the issue of the consultant's past history will likely be raised by the content owner in an effort to undermine the credibility of the system owner.

Invitations for hacker attacks

Some organizations now test the security of their products or systems by publicly inviting people to try to break through their security. A highly visible example of this process was used by the Secure Digital Music Initiative (SDMI). SDMI developed several technologies for on-line copyright protection, then invited the public to try to break those systems. SDMI offered an award of $10,000 to anyone who could defeat the security systems (http://www.hacksdmi.org).

From a technical perspective, this type of approach is quite sensible. From a legal perspective, before implementing such a public trial, consider the possible future implications of this type of program. If the technology tested by the program is ultimately put into widespread use and is eventually compromised, there is at least some risk that a party harmed by the breach could argue that your organization's decision to invite attacks unreasonably contributed to the eventual security breach that caused their injury. Invitations to hackers for test purposes can be effective and useful under some circumstances, but may not be worth the risk in others.

Information sharing

Organizations are generally reluctant to share information about network security breaches with other companies or with law-enforcement authorities. This reluctance is partially caused by concern that if the security breaches are widely known, more parties will try to attack the organization's network, thinking that it is vulnerable. Many industry experts believe that this reluctance to share information about security threats is inappropriate, suggesting instead that disclosure of those threats and attacks can improve security measures and reduce the risks for the future.

Law-enforcement authorities now have great interest in this subject and are urging greater cooperation in sharing knowledge of security threats. Some private parties fear, however, that mandatory disclosure of this information could increase the frequency of attacks and would provide more information to government authorities about their systems than they would like those authorities to possess. If voluntary information sharing does not increase, some advocate the implementation of laws or regulations specifically mandating disclosure of information on security problems. It seems far more likely that voluntary information sharing is to be preferred over mandatory disclosures, but failure to adopt voluntary measures may spur governments to force disclosure. Some industries, such as the financial services industry, have active and sophisticated practices for sharing information technology security knowledge and expertise. Those industries provide a very good model for the ways in which private, cooperative action to promote security information sharing can be effective and perhaps less intrusive than government-mandated actions.

There are already some circumstances in which disclosure of information regarding security breaches may be mandatory. For instance, companies that sell stock to the public in the United States may be required to disclose computer system security breaches to current and potential investors, to the extent that those security breaches are serious enough that they are considered a *material* factor in the decision-making process of the investors. Those disclosures most commonly take place in the mandatory information filings publicly traded companies in the United States make with the U.S. Securities and Exchange Commission.

4

Preventing System Misuse

This chapter discusses issues of legal rights and liabilities associated with misuse of computers and computer systems by authorized users. Previously, we addressed the challenges associated with managing access to computer networks and their content by unauthorized users. Here, we will examine the rights and liabilities to be considered when managing computer system use by authorized users. Effective computer network management requires both safeguards against access by outsiders who are not authorized to access the network and proper management of authorized users to ensure that those users stay within the scope of their authority.

Authorized users may include employees, contractors, business partners, or customers. System misuse by authorized users takes the form of either deliberate misconduct or inadvertent misuse, and it generally involves uses of the system that are beyond the scope of authority granted to a particular user. In such instances, system owners face potential legal liability from third parties and they possess certain legal rights enforceable against the authorized users. System owners must also be mindful of the fact that, under some circumstances, their efforts to manage authorized use, or their failure to manage such use effectively, may lead to legal liability to the authorized users.

The best way to prevent misuse of computer systems by authorized users, and minimize the risk of liability for such conduct, is to create and enforce effectively authorized use standards (also known as acceptable use standards). Those standards will help authorized users avoid inadvertent system misuse, and they will provide valid legal notice to system users of both the limit to user authority and the legal rights that the system owner is prepared to enforce against them. This chapter provides guidance as to appropriate basic principles for authorized use standards.

Liabilities caused by employees

Employees of businesses are agents of those businesses. As such, employees act as representatives of their employers, and the conduct of employees can lead to direct consequences for their employers. Employee use of enterprise computer systems can be a source of significant legal liability for employers. That liability can arise under several different fields of law, and it must be controlled through effective employee use policies and procedures. For example, an employee who makes improper use of computer software or other intellectual property that the company has licensed can create legal liability on the part of the company which is enforceable by the licensor of the property. Under such circumstances, the licensor would sue the company based on the conduct of the employee. If successful, damage awards would be assessed against the company, all based on the conduct of the employee. Employers must implement practices that enable them to manage computer use by their employees in order to reduce the risk that misuse of company computer resources could lead to legal liability for the employer. Those practices to curb misuse of computer networks by employees are also necessary to enable employers to preserve all of the legal rights available to those employers.

Employer liability to employees

In some instances, employers may bear legal liability to their employees, based on actions taken by the employer to manage or control employee use of company-owned computers or computer systems. Employers must be cautious to ensure that their efforts to reduce the risk of unauthorized system use do not result in potential legal liability to those authorized users. For example, employees may be able to raise privacy claims or breach of contract arguments based on usage-monitoring systems implemented by an employer to manage its computer operations. In these situations, the employer's efforts to control use of company computers and networks may create legal claims that employees can raise against their employer.

Employers may also bear liability to their employees for failure to manage company-owned computer networks effectively. In these cases, unauthorized use of the computer system may create claims enforceable by employees against their employer. Laws against discrimination or harassment provide an illustration of this concept, as they often permit employees to raise legal claims against their employers if those employers create a hostile work environment, or if the employer becomes aware that such an environment exists and does not effectively move to remedy the situation. Those legal requirements have been enforced in situations where a company computer system was used to create or distribute threatening or sexually oriented e-mail messages or Web content, for example. Claims of this sort are increasingly common and have been raised against a variety of organizations.

Employee liability to employers

Misuse of an employer-provided computer system by an employee can generally result in termination of employment, but it can also create legal liability on the part of the employee. That legal liability may involve a criminal penalty or a civil (or private) law claim by the employer. For instance, an employee who deliberately sabotages an employer's computer network can be held legally responsible for that conduct under contract law principles or under specific statutes, such as the CFAA at the federal level in the United States. Most of the criminal laws around the world that regulate unauthorized access to computers and their content, which were discussed previously in this book, are also applicable to authorized computer system users who exceed their usage authorizations. From a legal perspective, it is generally accurate to assume that an authorized computer user who uses the computer system in a manner outside of the scope of his or her authorization has now become an unauthorized user and can face legal prosecution. A rogue authorized user can, accordingly, be punished under criminal laws in most of the jurisdictions in which there are criminal statutes applicable to unauthorized access to computers or computer content.

Although most of the general laws prohibiting unauthorized access to computers or their content can clearly be interpreted to apply to authorized users who exceed their authority, even if those statutes do not expressly make that statement, some countries have clearly added that coverage to the text of their criminal statutes. For example, Article 55(b) of the Criminal Code of Belgium indicates that exceeding system use authority is a violation and is punishable by six months to two years in prison.

Employers should note, however, that many of the criminal laws applicable to unauthorized computer access make some requirements on the system owner before they will be applied. The most common of such owner requirements is the use of effective access controls and restrictions (e.g., encryption, passwords). Prosecutions under those laws can be successful only when the owner of the compromised system has satisfied the requirements imposed by the statute. For instance, Japan's Unauthorized Computer Access Law (Law No. 128, 1999) provides for fines and prison terms of up to one year for unauthorized access, but it also requires that the system that was compromised was a restricted use system, protected by access controls. Similarly, Finland's criminal laws (PENAL CODE ch. 38, § 8) provide protection against unauthorized access that involves bypass of a protective system.

Owners of computer systems should make sure that their systems qualify for the protection of local laws against unauthorized access and use by meeting all of the owner requirements imposed by those laws. Failure to do so could result in an inability to support effective prosecution of available rights. The most common such requirement, as illustrated above, is the requirement that the system be protected by reasonable security measures. The Computer Misuse Act of 1990, enacted in the United Kingdom, presents another example of an owner requirement. That statute requires that the defendant knew or should have known, at the time of the unauthorized computer access, that the access in question was not authorized. This type of requirement suggests that owners of computer networks must make sure that all authorized users understand the scope and limits of their authority. If an authorized user does not understand the limits of his or her authority, and if that lack of understanding is caused, in whole or in part by actions (or inaction) on the part of the network owner, that owner may not be able to pursue all potential legal rights under criminal laws such as those in the United Kingdom.

This discussion should not be viewed as advocacy in support of prosecuting employees and other authorized system users under criminal laws against unauthorized access in every instance in which they act outside of the scope of their authority. These prosecutions are very serious actions and should not be taken lightly by the system owner. It is, however, important for those owners to be aware of both the potential rights they have under the criminal laws, and the operational requirements they may face if they want to preserve the ability to exercise those rights.

Civil or private law claims can also be raised by employers against employees, based on employee misuse of employer-owned computers. If the employee works under a contract with the employer, the employer can often sue the employee for breach of the employment contract. For example, if an employee misuses the company's computer to access or distribute trade secrets owned by the company, the employer would likely be able to terminate the employee's contract and recover monetary damages under a contract law claim. To preserve this legal right, employers who enter into contracts with their employees should make sure that all such employment agreements clearly indicate that the employer has the right to terminate the contract in the event of computer misuse by the employee, and that the employee will be liable for all damages resulting from the system abuse. Even when employers do not use contracts with employees (e.g., in jurisdictions in which employment arrangements are deemed to be "at will," that is, either party can terminate the arrangement at any time), employers should put all employees on notice (in employment handbooks, employee manuals, and direct notices) that the employer reserves the right to terminate employment based on computer abuse and that the employer also reserves the right to pursue all available legal remedies against an employee who engages in such conduct.

In some jurisdictions, these private law claims may also include lawsuits to recover damages for a breach of a legal duty owed by the employee to the employer. These claims are in addition to any contract claims that the employer may possess. In the United States, for instance, certain tort law claims are available to employers if the employee's misuse of the computer system resulted in quantifiable damages to the commercial interests of the employer. For instance, if an employee used the company's computer system to distribute e-mail messages that harmed the commercial reputation of the company, the company could seek monetary compensation from the employee, in some jurisdictions under a tort law theory, claiming a form of defamation. Similarly, if an employee's misuse of a computer system resulted in damage to the system (e.g., introduction of a virus, damage to data stored in the system), the employer would likely be able to pursue a tort claim against the employee to recover compensation for the damages resulting from that misuse.

To recover under tort claims, an employer will be required to prove several facts. It will be required to prove that the employee had a duty of care with regard to the use of the computer system. The employer will also have to show that the duty of care was actually breached by the employee. Finally, the employer must demonstrate that the breach of the duty by the employee caused damage to the employer and the employer must be able to quantify the amount of those damages.

Liability associated with other classes of system operators and users

Liability issues are also present in the context of computer systems that are made available to the public for use on a commercial basis (e.g., ISP network, Web-content hosts, application service providers, and other on-line service providers). In this context, the system owner is not an employer and the system users are not employees. Instead, the system owner is a commercial service provider and the authorized users are customers of that service provider. These system operators and their customers are potentially liable to each other, or to third parties, based on the use of those systems. For example, an ISP can be held legally liable for copyright infringement or defamation as a result of material that its customers distribute on its network. A system operator that opens its network to commercial users can be liable to, or raise claims against its customers based on unauthorized network use. Third parties may raise claims against the system operator based on unauthorized use of the network by the customers.

Some jurisdictions provide at least partial protection for on-line service providers and other computer system operators who handle public data traffic. Federal law in the United States, for example, protects on-line service providers from liability for content provided by third parties that is distributed using their networks, if those on-line service providers take reasonable steps to manage that content and to respond to claims of improper content. The Digital Millennium Copyright Act in the United States generally frees on-line service providers from liability for content made available by other parties using the networks of the service providers if the service providers implement procedures to be used to remove unauthorized content from their networks swiftly and effectively, when they are presented with reasonable evidence that the content in question is either inappropriate or its on-line use has not been validly authorized. Note that the protection provided to the on-line service providers is not absolute. To qualify for that protection, the service provider must present a clear and effective system that third parties claiming inappropriate use of the service (e.g., distribution of defamatory or infringing material) can use to persuade the service provider to review, and if appropriate, remove the material in question, or discontinue service to a user who is abusing the service.

The diversity of on-line service providers continues to increase dramatically. For example, on-line auction service providers are now the center of significant attention with regard to unauthorized use. It remains an open issue to what extent an on-line auction operator can be held legally accountable for products sold using its system. We will address the issue of potential legal liability for on-line auction operators again later in this book. At present, most on-line auction operators take the general position that they are sales conduits and transaction facilitators who should not be held responsible for the conduct of the buyers and sellers who use their systems. For instance, the on-line auction operator would generally contend that if a seller using its service conducted a fraudulent sale, it would work with the defrauded buyer and law-enforcement authorities to try to remedy the situation, but the auction operator itself should not bear any responsibility for the misconduct of the system user.

French authorities did not accept that argument, however, in a case in which Yahoo! was a party. The *Yahoo!* case involved the sale of goods that were illegal in France, to a French buyer, by a third-party seller using the Yahoo! on-line auction process. The French court applied French law and found Yahoo! liable for the transaction. The court ordered Yahoo! to block all such transactions in the future, and Yahoo! is contesting the enforcement of the French court order in a U.S. court; however, the French ruling still stands, and the case serves as a warning to on-line sellers in general, including auction operators.

In spite of the *Yahoo!* case, the precise extent to which an on-line auction operator may be legally liable for misconduct by sellers who use the site remains unclear in most jurisdictions. At present operators of these systems should not assume that the general argument that they are merely conduits who are not responsible for the transactions they process will be accepted in all jurisdictions. This issue is discussed more fully in Chapter 8, but for current purposes, recognize that it is likely that on-line auction operators will face a significant risk of some level of responsibility for misconduct by users of their systems.

On-line service providers can look to contract law as a source of legal rights enforceable against the users of their systems. Service agreements, terms of use, and other forms of contracts are commonly used by service providers to establish the rights and obligations of the service provider and the user. In effect, those agreements define the scope of the user authorization. User conduct that varies, in a material way, from the user obligations established in these contracts, provides the basis for legal claims of breach of contract by the service provider against the user. On-line service providers should, accordingly, ensure that they enter into binding contracts with their customers and that those contracts adequately describe the acceptable use standards for all users of the service.

On-line service providers can also look to the criminal law provisions described previously to support claims against commercial users who abuse their access to the services. Just as an employer can make use of the criminal sanctions against computer abuse in the event of an employee's misuse of its computer network, so too can a commercial on-line service provider apply those laws against a customer who misuses the service. Criminal sanctions applied to inappropriate access to computers or their content are applicable to users of commercial computer services. For example, if a customer distributes illegal content using an on-line service, that user can face criminal prosecution for both the illegal content and for the abuse of the computer service (i.e., for using the computer system to engage in the illegal conduct).

Harassment

At the federal and state levels in the United States, it is illegal to create a work environment in which employees feel threatened or harassed on the basis of gender, religion, race, or ethnic origin. It is important to make sure that your organization's computer and telecommunications systems are not used in a manner that helps to create an illegal threatening environment for any of its employees. For example, e-mail messages created or distributed using your company computer system that are harassing or threatening in nature can create legal liability for the employees who create or distribute those messages and for the company itself. Employee access to Web content that is inappropriate for the workplace (e.g., sexually oriented material) can also create liability for the individual worker and for the company.

Harassment claims are raised by employees who are members of the protected classes and who feel threatened by the conduct. For example, an employee who is a member of a particular race or ethnic group may raise legal claims if he or she can demonstrate that the employer has created, or has permitted, a work environment that is hostile to the employee because of the employee's race or ethnic origin. These legal claims involve lawsuits against individual employees who participate in the harassment or discrimination, and against the employer who permitted the conduct. In today's workplace, a major part of the work environment involves some form of computer use, and this is how computer network misuse can become a key part of an employee's claim of harassment. Even if the employer did not support or cause the illegal harassment, that employer can still be held legally liable if it did not take effective measures to prevent the harassment. An employer that permits its computer system to be used for these illegal activities can bear civil law liability to employees who are affected by the illegal conduct.

When harassment takes the form of threats of physical harm, it is commonly characterized as stalking. Various jurisdictions now specifically prohibit stalking. In some instances, computer systems and communications (e.g., e-mail) are used as part of the intimidation campaign that is commonly associated with stalking. If an organization becomes aware that its computer system is being used to facilitate stalking, the organization has a legal duty to report that situation to law-enforcement authorities and to cooperate with the efforts of those authorities to resolve the problem. Failure to assist the authorities in those cases can result in legal liability for the company.

Defamation

In many jurisdictions, it is illegal to publish false statements about a person or organization that result in harm to the reputation of that party. The party who is the subject of the false statements can sue the publishers of the statement for monetary damages. Any party who publishes a false statement that results in such harm can be held liable for the damages that are incurred as a result of that publication. Accordingly, if an employee of your organization uses the company computer network to distribute a defamatory statement, the party who is the subject of the statement could seek damages from both the individual employee and your company, as both the employee and the company would be viewed as publishers of the false and harmful statement. Damages for defamation are commonly awarded to compensate the injured party for the harm to its reputation caused by the publication of the false material.

In many jurisdictions, defamation claims can be raised by both individual people and businesses. For example, if a computer system user distributes e-mail messages that defame the management of a business, the individual managers in the organization can sue, and the company itself can raise defamation claims. The law in many jurisdictions establishes defamation as a legal cause of action for both people and organizations that have legal status.

Willingness to enforce defamation claims varies widely among jurisdictions. For example, defamation claims are generally difficult to win in the United States, largely because of its heritage favoring widespread freedom of expression. Other jurisdictions, such as the United Kingdom, however, are far more active in their enforcement of legal protection against defamation. Computer network operators should be aware of their potential exposure to defamation claims, even in those instances in which the defamatory publication processed by their network was not created by the operator and was not authorized by the operator.

Protection against defamation liability should begin with the authorized use standards implemented for each computer network. Those standards should provide a brief definition of defamation and they should clearly indicate that defamatory material is prohibited from the system. The standards should indicate that the system will be monitored to prevent defamatory use, and that the system owner reserves the right to remove all material that it deems to be potentially defamatory from the system.

Financial disclosures

Publicly owned corporations in the United States must control the information they release regarding their financial condition and activities. The leading enforcer of that obligation is the federal government agency, the SEC. Failure to comply with the appropriate financial disclosure obligations can result in legal penalties for both individuals and the company itself. Generally, there are two types financial disclosure obligations. One involves mandatory disclosure of information that is relevant to the investment decisions of current and potential investors (i.e., information that is "material" to the investment decision-making process). The other type of obligation is one that limits disclosure of financial information at certain times (e.g., the so-called "quiet periods" immediately prior to public stock offerings).

As computer systems are now commonly used in the distribution mechanism for financial information (e.g., through e-mail, Web publications), employee decisions and actions regarding electronic distribution of information regulated by the securities laws must be effectively managed. Failure to exercise effective management in this context can result in a failure to disclose information that should have been disclosed or a disclosure of information that should not have been disclosed. In both instances, the failure could lead to personal liability for the employee involved and other liability for the company. Those legal claims could be asserted by the SEC, state regulators, or by private parties harmed by the violation (e.g., shareholders of the company).

One example of this type of unlawful disclosure can involve release of material financial information prior to the appropriate time or to individuals who should not receive the information. During a quiet period when management of a publicly traded company is prohibited from public release of financial information pertaining to an upcoming stock issue—for instance, if a company's employee were to release an informational e-mail intended only for internal distribution—that release could constitute a public disclosure in violation of the securities laws.

Other nations are beginning to consider adopting laws regulating disclosure of financial information to investors. Businesses that solicit outside investors will find an increasingly broad set of information disclosure rules as time passes. Those rules present an important source of potential liability associated with computer system misuse, and all authorized use standards developed by companies that solicit the public for investment should address those financial disclosure requirements.

Personal data

Although the issues of data security and information privacy are addressed in greater detail elsewhere in this book, it is important to recognize that misuse of personal information by authorized users of computer systems can lead to legal liability for both the individual system user and for the system owner. If your organization collects personal information from the public, special safeguards should be implemented to make sure that the information is not misused. Legal requirements associated with management of personal information are of significant importance. In the context of this chapter, the most important point regarding data privacy is that unauthorized use of computer systems can lead to violations of the data privacy obligations. Those violations can result in legal liability for the individual people involved in the data misuse and for the owner of the computer system involved in the misuse.

For example, the EC has established laws governing the collection and use of personal information, through its information privacy directives. Each of the member nations has enacted national laws to implement those directives. In the United Kingdom, for instance, the Data Protection Act brings the EC privacy directives into effect. Violations of the terms of the Data Protection Act can result in civil and criminal penalties. Those penalties can be directed against companies and individual people. Thus in Europe, misuse of a computer system that results in violations of the information privacy laws can lead to legal liability for the individual people involved in the misuse and for the owner of the system involved in the privacy violation. Standards of use for computer systems should thus be coordinated with the information privacy practices of the organization to make sure that the usage standards comply with all relevant information privacy requirements.

Intellectual property

Potential legal liability associated with improper use of intellectual property is addressed in more detail elsewhere in this book, yet it also merits brief discussion here. If your enterprise makes use of intellectual property created and owned by other parties, that material must be used in a manner consistent with the license agreements into which you have entered. When your computer system is used to store, distribute, or use licensed intellectual property (e.g., computer software), you must ensure that all system users comply with the appropriate license terms. You must make sure that the authorized users of your system abide by all intellectual property requirements as to the content they distribute on the system. Failure to comply with those terms can result in liability to the owner of the intellectual property for claims such as copyright infringement.

For example, if your company has properly licensed software from a vendor, but one of your employees makes a copy of the software, in violation of the terms of the license, your company will, in most cases, be responsible for that breach of the license. To expand on this example a bit, imagine that you are an applications service provider, hosting various types of applications software, much of which you have licensed from other parties. Now imagine that some of the clients who access software through your service violate the terms of that access. Such a situation would create potential intellectual property law liability for the client and for your business.

Effective management of computer system use is also essential to protect intellectual property that the system owner claims as its property. For example, a company that develops software or on-line content commonly relies on its employees to create that property. The legal principle that enables an employer to assert ownership over intellectual property created by its employees, within the scope of their employment (i.e., as part of their regular job assignments), is the concept of *work-for-hire*. Disputes sometimes arise over the ownership of intellectual property created by employees when employees try to claim ownership over their work based on the argument that the work was created outside of the scope of their employment. The best way to reduce the risk of encountering this type of dispute is to implement clear and consistent computer system use standards. Properly structured and enforced, those standards can help to ensure that an organization's computer assets are used only for purposes authorized by the organization, thus substantially reducing the chances of intellectual property ownership disputes with employees.

Misuse of intellectual property is also a significant problem for on-line service providers. For example, many users of publicly accessible commercial computer systems distribute copyrighted material in unauthorized ways. One clear illustration of this problem is the dispute associated with on-line music distribution using the Napster system. As the *Napster* case in the U.S. courts demonstrates, for instances in which users of a commercial on-line service violate intellectual property use restrictions, the provider of the on-line service may face significant legal liability, based on the inappropriate conduct of the users of its service. Rather than pursuing claims against millions of individual people, as the music copyright owners could have done in the *Napster* case, those owners chose to focus their legal action against the service provider, Napster. This tactical approach is commonly applied, and it is a major reason why computer system operators must recognize their potential liability based on the conduct of their customers.

The *Napster* controversy also provides a hint of future difficulties associated with management of authorized users in a computing environment that is likely to rely increasingly on distributed-computing models that facilitate widespread digital content sharing. Peer-to-peer content sharing systems that are even less hierarchical in structure than Napster (Gnutella and Scour systems, for instance) will provide a significant challenge for authorized use management. As more network users obtain greater control over the content access decisions, we are likely to see more instances of unauthorized access.

In a distributed-computing environment, any single user can make content widely available quickly and without immediate detection. Yet even in a distributed-computing environment, parties who license intellectual property will be held accountable for management of that property in a manner consistent with their authorization. If a company licenses software, for example, and wants to permit its employees to access that software using a peer-to-peer sharing system, the company will be held responsible for managing that sharing system in a manner that ensures that all shared use complies with the terms of the software license. Failure to manage the shared use effectively could result in liability for the company. In such a setting, it is likely to be increasingly difficult to manage the conduct of authorized system users, and enterprises that make use of the content-sharing models may face a greater risk of liability for intellectual property misuse.

Trade secrets

Commercial information that provides a company a competitive economic advantage over its rivals is protected from unauthorized disclosure or use in most jurisdictions under either civil or criminal law. The valuable, secret information is commonly described as trade secrets. If authorized computer system users misuse trade secrets, that conduct can generate liability for the user and for the system owner.

An authorized user who misuses trade secrets can be liable to the owner of the secrets under a variety of legal theories. If that user had a contract relationship with the owner of the secrets (e.g., a confidentiality or nondisclosure agreement, an employment agreement, an authorized use agreement for access to the computer system) the user would likely be subject to a civil law claim of breach of contract. Some jurisdictions (e.g., various states in the United States) also provide a legal claim, under tort law principles, for misappropriation of trade secrets.

In addition to civil or private law claims, many jurisdictions also provide criminal law sanctions for theft or misuse of trade secrets. For example, the United States, under its Economic Espionage Act (18 U. S. C., §§ 1831–39) makes it a federal offense to access trade secrets for economic espionage purposes. Unauthorized access to trade secrets through use of a computer system constitutes unauthorized computer access, and is thus prohibited by the criminal laws that protect computer equipment and content from unauthorized use. In addition to that general protection, several nations specifically prohibit unauthorized access to trade secrets through computer systems. For example, Portugal's Criminal Information Law (CRIM. CODE ch. 1, art. 7) provides for prison terms of one to five years if there is unauthorized computer use that results in unauthorized access to trade secrets. Denmark's PENAL CODE § 263 imposes jail terms of up to two years when computer system misuse involves an attempt to steal trade secrets. Similarly, Germany's prohibitions against data espionage (German PENAL CODE § 202a) provide for prison terms of up to three years when computer misuse involves data espionage activities.

A system owner can also face legal liability for misuse of trade secrets, even if the owner did not directly engage in the misconduct. For example, if a company has received authorization from the owner of trade secrets to make certain use of those secrets, the company would likely bear legal liability if an employee or contractor of the company violated the terms of the authorization. That liability could arise under a claim of breach of contract or a tort claim of misappropriation of trade secrets. Under either theory, the owner of the trade secrets would likely sue both the individual person who misused the secrets and the company that had been granted access to the secrets.

Consider a situation where company A enters into a contract with company B, and that contract provides that company B will create and manage a database containing sensitive information about company A's customers. If an employee or contractor of company B were to disclose some of that information to another party, in violation of the terms of the contract, company A would likely go to court against company B, seeking compensation for economic damage resulting from the unauthorized use of that information.

Misuse of trade secrets owned by another party can also raise potential criminal liability for the owner of the computer system, even if the system owner was not directly responsible for the misuse. Most criminal laws that prohibit unauthorized access to computer content provide penalties for both the party who actually engaged in the unauthorized access and for parties who facilitated that misconduct. If prosecutors can show that the system owner participated in the misconduct, that owner can bear criminal liability for contributing to the misconduct. Even if the system owner did not authorize or otherwise directly participate in the misconduct, the owner may be criminally liable if he or she were negligent in his or her management of the people who actually engaged in the misconduct. Accordingly, in the example given above, if prosecutors can demonstrate that company B either directly participated in the misuse of the trade secrets or it contributed to the misuse through negligent conduct (such as poor oversight of the computer system and its users), company B and its management could face criminal prosecution.

The best protection against the loss of your own trade secrets and the potential liability for disclosure of the trade secrets of others is effective management of authorized use standards. All trade secrets and other commercially sensitive material should receive the highest level of usage control and monitoring. If your organization implements effective usage controls, the chances of trade secrets abuse decrease, and your ability to defend your organization against legal claims in the event of inadvertent disclosure increases.

Export Controls

Computer systems often store and process material that cannot be freely distributed to parties in other countries or to foreign citizens. A common example of one form of regulated content is encryption software. Export controls in the United States (e.g., the Export Administration Regulations, 15 C. F. R. § 768) and other countries are primarily designed to regulate the international distribution of goods, services, and information that can have military applications. These rules generally require government approval (e.g., export licenses) prior to the international distribution of the controlled material.

Failure to obtain a license prior to export when required, can result in severe penalties that include both fines and prison terms. Fines can be applied to companies and to individual employees, and prison terms can be assessed against the responsible employees. Computer systems are commonly used to distribute information internationally, and in that capacity, their use must be monitored effectively to ensure that material within the scope of export control regulations is not circulated before necessary government authorizations are obtained. If an employee of your company violates export control restrictions, the employee may be personally liable for the violation and your company will, in most cases, be held responsible as well.

Computer system content that is subject to export regulations must be managed carefully to ensure compliance. This can be a particularly great challenge as export restrictions often apply to goods, services, and information. Export control rules in the United States, for instance, apply to export of certain forms of encryption hardware and software, but they also apply to international distribution of information that could help a foreign party to develop or manufacture encryption systems that fall within the scope of the regulations. With this broad scope, the U.S. rules affect not only the international distribution of actual encryption products, but also the international delivery of support services and background information associated with those encryption products. While management of international distribution of goods and services can generally be accomplished effectively, many organizations find that control over distribution of information is very difficult. Authorized use standards should, accordingly, address the issue of controlling access to content that falls within the scope of those regulations. Failure to do so effectively can result in inappropriate system users accessing the restricted content, and legal liability for both the individual users and the system owner can result.

Antitrust and competition law

Use of an enterprise's computer system to engage in antitrust or competition law violations can create liability for the individual user and for the enterprise. These laws are designed to promote fair commercial competition and are in effect in many nations throughout the world. Examples of violations of these laws include collusion among competitors aimed at reducing competition, price fixing, unreasonable product tying arrangements, and unfair or deceptive trade practices. Often times, communications between businesses provide the core of evidence in support of antitrust or competition law cases. Those communications (e.g., e-mail) sometimes, for example, provide evidence of collusion to divide up markets, to fix prices or to act in some other manner that threatens competition. Businesses must act to make sure that their employees and other agents do not use the computer system to conduct activities that violate antitrust or competition laws.

A highly visible example of antitrust litigation was provided by the U.S. government's prosecution of Microsoft Corporation. A trial court found Microsoft to be in violation of U.S. antitrust laws, but that decision is on appeal at the time of this writing. A highly publicized aspect of that trial involved disclosure to the court of e-mail messages and other electronic documents that supported the government's allegations of illegal conduct. It is likely that many organizations around the world have, in their electronic records, documents that could be used in a legal proceeding to support allegations of antitrust or competition law violations. Businesses should strive to ensure that their computer systems are not used as either instruments to engage in antitrust law violations or as records repositories that document that conduct.

Various jurisdictions around the world, in addition to the United States, are increasingly attentive to competition law concerns. The European Union, for example, has been highly active in monitoring commercial conduct from an antitrust perspective. The Japanese are increasingly active on antitrust and competition law matters. As the *Microsoft* case illustrates, evidence of violations of competition law can often be found in the e-mail and other computer records of businesses. In that environment, it is particularly important that businesses develop and enforce standards of use for their computer systems to reduce the risk that their employees may make use of those systems to engage in activities that violate antitrust or competition laws.

Acceptable use standards

Employee conduct can damage an organization's computer system and it can cause legal liability for that organization. Accordingly, enterprises that operate computer equipment and systems should create and enforce standards for acceptable use of those systems. For greatest effect, these standards should be coordinated with the computer access, network security, and information privacy policies and practices of the organization.

One of the challenging issues associated with acceptable use standards is the question of personal use of company-owned computers. Personal use may take the form of sending or receiving personal e-mail from a computer at the office, browsing Web sites at work for personal reasons or making personal e-commerce purchases while at work. Common practice is for businesses to prohibit personal use of company-provided computer systems for personal activities. Effective acceptable use standards should clearly indicate that the company-owned computer system is to be used only for business purposes.

In spite of these prohibitions against personal use of computer systems, however, many organizations choose not to enforce the personal use restrictions rigorously. This approach, often pursued as some level of personal use of company computers, is viewed by many people to be an important employee benefit that helps to improve morale and make the organization a more attractive place to work. In many instances this approach is perfectly sensible for business reasons, but if your organization adopts such an approach, you should be aware that the strategy might undermine your ability to enforce the prohibition aggressively in instances when you believe such enforcement is necessary. Inconsistent enforcement of a personal use prohibition can also lead to additional legal liability for the company. For example, allegations of discrimination or unfair labor practices may arise when an employer enforces strict acceptable use standards for some employees but not for others. As a general matter, it is better to avoid inconsistent applications of the standards, but if such inconsistencies of enforcement can be justified for legitimate business or technical reasons, the system owner can more readily defend them if the enforcement practices are ever challenged.

Some organizations have attempted to extend the reach of their control over personal computer use by their employees into off-site activities. This extended reach raises serious legal issues of possible violations of the rights of employees. For example, in the United States some employers have fired employees allegedly based on on-line activities conducted by those employees during their off-work hours. In several different ongoing cases, it is reported that employers who became aware that some of their employees were operating on-line adult-oriented businesses during their free time fired those employees, allegedly based in part on their on-line activities. Clearly an employer has the ability to terminate employment based on misuse of company-owned computer systems; however, these cases seem to involve purely personal computer use outside of the legitimate scope of employer control.

Another growing area of confusion involves the increased use of off-site computer equipment and network access provided by employers for employees. Many companies now provide, at company expense, personal computers, laptop computers, and Internet access for use by their employees at their residences or other locations that are not part of the employer's property. One open issue associated with this development is the question of the extent to which acceptable use standards can and should be applied to use of that off-site networking capability.

Certainly an employer has legitimate interests in ensuring that the off-site computer system access it provides to its employees and contractors is managed in ways that serve the business interests of the employer and protect the security of the employer's property. For example, authorized use guidelines that apply to security measures to be used for company-owned equipment and information while located off company premises are both appropriate and necessary. Similarly, security requirements associated with procedures to be applied to off-site access, via public communications networks, to the company computer network are well within the scope of the employer's legitimate business interests and should be included in the acceptable use standards.

In contrast, however, usage standards that attempt to govern personal use of the off-site equipment during nonworking hours should be approached by employers with great caution. If a decision is made by the employer to provide off-site computer equipment and network access, the employer should be prepared to permit personal use of that equipment and access by employees. Standards of use, in that context, should be confined to those activities that have a direct connection with legitimate efforts by the employer to manage the security of its computer and information assets. If an employer is uncomfortable limiting its control over use in that setting, he or she should probably reconsider whether provision of that off-site equipment is in the company's best interest. To maximize its ability to manage computer system use, an employer should keep access to the system confined to work hours and company premises. When an enterprise makes the business judgment that it wants to extend network access to off-site locations, it should recognize that its system management capabilities will invariably be diluted and made more complicated.

Acceptable use standards should explicitly indicate that the company-owned computer system is not to be used for any activity that violates any law or regulation. The standards should also note that the system is not to be used in a way that violates the policies and practices of the company or infringes on the legal rights of any other party. These broad standards should be clearly expressed to all system users in written form, and the system owner should take all reasonable measures to ensure that the system users in fact understand the type of conduct that may be illegal. This discussion should include at least a brief summary of the major legal obligations associated with authorized use. The legal obligations discussed should address at least the following topics: (1) harassment/discrimination, (2) intellectual property and trade secrets, (3) information privacy, (4) financial disclosure/reporting requirements, (5) defamation, (6) antitrust and competition law, and (7) export controls.

Acceptable use standards should clearly identify the potential penalties associated with violation of those standards. They should clearly state that violations of the acceptable use standards are cause for termination of employment. The standards should also state that the company reserves the right to seek criminal prosecution or pursue civil (private) legal claims against an employee or former employee based on violations of the standards.

The standards should also establish an obligation on system users to report suspected violations of the standards. It should be clear that this reporting requirement places an obligation on users to report incidents in which they may have inadvertently acted outside of the scope of their authority. The document should describe how and to whom those suspected violations should be reported. It is often helpful if the standards also provide users with some sense of what will happen after they report violations. Users should be reminded that failure to report violations is, in itself, a violation of the acceptable use standards.

Acceptable use standards should explain the obligations of the employee that are effective upon termination of employment and after leaving the organization. These obligations generally include surrendering all company-owned equipment and content in the possession of the employee and cooperating with the company to ensure that all authorized use privileges of the departing employee are effectively terminated.

Acceptable use standards will not prevent all computer misuse and they will not provide total immunity from legal liability for the computer system owner. The standards will, however, significantly reduce the risk of inadvertent system abuse. The standards will also help the system owner to reduce the scope of its liability in the event of an incident. Standards that are effectively enforced help to demonstrate good faith, reasonable efforts at prevention by the system owner. In addition, they serve to define the scope of authorization for system users and thus help the system owner to argue successfully that any actions inconsistent with that definition of authority represent conduct beyond the scope of authority granted by the system owner. This argument helps to reduce the risk of owner liability by undermining suggestions that the misconduct was executed by the user in the capacity of an authorized agent of the system owner.

Monitoring of employees

Computer system usage by all authorized users should be carefully managed. System usage monitoring can be an important element of those management efforts. Employers have the right to monitor the ways in which their employees use computer systems that are provided by the employer. This right is very broad, but it does not permit the employer total discretion as to monitoring. In many jurisdictions, including the United States, essentially, all monitoring activities that are reasonably related to an employer's efforts to protect its property from damage or loss (including both tangible property and intangible property); to protect its legitimate business interests; or to protect itself from potential legal liability are permissible. Two sets of laws that commonly restrict that employer flexibility, in some jurisdictions, are privacy regulations and labor or employment laws.

With regard to privacy laws, jurisdictions that restrict the ability to gather, distribute or use personally identifiable information (e.g., telephone numbers, mailing addresses), such as the EC, indirectly regulate some aspects of the ability of employers to monitor employee use of computer systems. For example, while those laws would certainly not prevent an employer from monitoring employee use of the company-owned computer system, they would likely make it necessary for the employer to provide notice of that monitoring to the employee and to provide the employee with the opportunity to review the personal information collected through the monitoring. If an employer monitored e-mail message is sent and received by the employee, the employer would likely have a legal obligation, under the European privacy laws, to provide prior notice of the monitoring to the employee and an opportunity for the employee to review the information the employer collects.

When considering implementation of an employee-monitoring system, employers should consider the following factors. They should identify the specific property, business or legal interests that the monitoring is intended to protect. They should consider the extent to which the employee has a reasonable expectation of privacy as to the conduct that is to be monitored. Finally, employers should determine whether there is an alternative method to protect its legitimate interests that is less intrusive into the privacy of the employees. If employers conclude that the interest being protected is significant, that the employee does not have a reasonable expectation of privacy regarding the conduct in question, and that no other less intrusive option is reasonably available, then the employers should proceed with the planned monitoring system.

Remember that monitoring creates records and that the organization will be held accountable for those records. This issue was discussed in greater detail in Chapter 2, which addresses management of electronic records, but it is worth repeating the point here. If your company monitors employee computer use, those records will be accessible, through the discovery process associated with litigation, to law-enforcement authorities, and to private parties who sue you. If those records contain evidence of illegal activity, your organization will likely have a legal duty to report the illegal conduct and to cooperate with law-enforcement authorities to remedy the problem.

Once you have records in your possession, your organization will be assumed to have had actual knowledge of everything that those records reflect. For example, if an employee of your company makes use of the company computer system to commit fraud, and if your monitoring system obtains evidence of that illegal activity, the authorities will attribute knowledge of the illegal conduct to your organization, regardless of whether you actually reviewed the monitoring data and identified the illegal behavior. The lesson from all of this is simple. Before you implement an employee-monitoring system, consider the potential consequences of the knowledge that such systems may yield. If you choose to implement a monitoring system, make sure that you, in fact, review and effectively manage the data generated by the monitoring.

Monitoring of employee use of company computer systems is likely to raise employment and labor law issues in the future. Although they are only beginning to develop at the time of this writing, some issues are likely to arise with respect to future computer system monitoring in the context of employee or union rights. Issues including the extent to which employers can monitor the use of company-owned computer networks to restrict activities such as labor union organizing are beginning to emerge in some jurisdictions, including the United States. We may, at some point, see employer computer system monitoring practices addressed in union negotiations with management in an effort to have those practices addressed in collective bargaining agreements. The monitoring activities may also eventually draw attention from government regulators in the context of rules for appropriate labor relations and employer practices. Employers would be well advised to pay particular attention to the evolving state of regulatory and public policy focus on the topic of monitoring of employee use of computer systems.

Due diligence in hiring, training, and termination

Effective management of employee use of computers systems begins at hiring and is supported by thorough training practices. By hiring trustworthy and competent employees, the organization takes a major step toward protecting its computer assets from misuse. This process involves applying prudent caution (due diligence) to the selection and training of employees and other authorized system users. Time and effort invested in these human resource activities can yield significant dividends with respect to overall computer security.

Basic hiring practices such as thorough reference and prior employer checks prior to hiring should be a standard procedure for employers. For particularly sensitive positions, criminal record checks are useful and should become part of the hiring process. As a general rule, all educational transcripts and other forms of documentation used by job candidates to support their applications should be verified prior to the issuance of an employment offer.

Training in regard to the organization's computer system standards of use is important. That training should be provided at the time of hiring, and it should also include periodic refresher sessions to remind employees of their obligations and to apprise them of changes in the standards that will certainly be implemented frequently. Training is particularly important in support of effective computer system management for two reasons. For one, it helps to reduce the risk of inadvertent system misuse by making employees more adept at operating the system and understanding the limits of their authority. Training also helps to make users aware of the potential threats to the computer network, which can make them more useful in the effort to implement security measures.

When an employee terminates employment, basic procedures to protect the computer system and its content should be applied. Those procedures should include prompt deactivation of access mechanisms (e.g., passwords, identification cards) and retrieval of all computer equipment and digital content in the possession of the departing employee. At the time of termination (perhaps during the exit interview) all departing employees should be reminded of their continuing obligations to protect trade secrets and proprietary information, and to avoid conduct that compromises the security of the organization's computer operations. The focus of this process should be to remind the departing employee of his or her continuing legal obligations to the former employer.

Contractors and business partners

The competitive requirements of today's marketplace make increasing integration of company computer systems operations with those of contractors and other businesses a common practice. That practice raises special challenges as to effective system management by increasing the number of authorized users of the network. It becomes more difficult for an organization to manage authorized system users when the roster of those users increases in number and when some of those authorized users are employees of other companies.

Many of the same legal rights and responsibilities associated with management of employee use of computer systems also apply to authorized commercial users who are not employees (e.g., contractors). In some ways, however, management of these nonemployee authorized users can be more difficult for the system owner, as that owner will have less control over their conduct. From a legal perspective, the key mechanism to foster effective authorized use management in an environment of interconnected networks is the development of appropriate contracts between the parties to ensure proper coordination of their authorized use standards. In addition, the standards themselves must be compatible in order to reduce the risk of inconsistencies which will likely invite operational confusion and increase the chances of inadvertent system misuse.

Also significant when organizations are integrating their computer network operations are the issues of coordinating responses to incidents and sharing of liability when an incident occurs. The parties should address these issues before they actually interconnect their networks. The terms under which the two organizations will coordinate the response to any instance of computer misuse and the degree to which they will share liability resulting from that misuse should be critical factors in the decision-making process associated with the joint effort. If the parties cannot agree on how to coordinate emergency responses and liability sharing, they should not integrate their computer operations. Joint efforts undertaken without agreement on these key issues are doomed to failure.

Customer use

Many businesses today provide unprecedented access to their computer systems for their customers. This includes both individual customers and entire businesses that are customers in a business-to-business transaction. Consider for example, a financial services company. It will most likely permit noncustomers to have basic access to its Web site, retail customers to access secure on-line resources for account information and transactions processing, and business customers to purchase market analyses or obtain other customized wholesale financial services. That tiered approach to network access helps businesses to compete more effectively by better serving the different demands of different groups of customers. It also places significant burdens on computer system management efforts. If you think it is difficult to effectively manage authorized users who are employees, contractors, or business partners, those challenges are relatively simple compared to those facing enterprises that authorize their customers to have some form of access to the enterprise's computer network.

In this environment of ever-greater incentives to provide more customers with access to commercial computer systems, it becomes increasingly difficult to manage authorized use by those customers. The normal setting now assumes that many different types of customers with different levels of authorization will be using a single computer network. To manage that structure effectively, system operators must establish clear definitions of the scope of each user's authority, and the operator must devise practices and procedures that permit effective enforcement of those limits on user authority. The legal starting point for that entire process is contract law. Clear, accurate, and binding contracts must be established with all customers to govern their authorized use of the company's computer resources. Those contracts must effectively define the authority of the users and establish the legal mechanism through which the limits on each user's authority will be enforced. Traditional hard-copy agreements and electronic agreements are both effective and appropriate, and they should be considered the key component of effective management of customer access to commercial computer resources.

Deliberate misuse versus inadvertent conduct

Computer system misuse raises legal problems for individuals and enterprises regardless of whether the misuse was intentional or accidental. Standards of acceptable use for computer operations should be aimed at reducing both intentional and unintentional misuse. In most cases, the mere fact that computer misuse was not intended will not relieve the parties involved from all liability for the consequences of that misuse. Proof that the illegal conduct was not deliberate may help to reduce the penalty associated with the conduct, but it will generally not provide a complete defense for the action.

Given the need to prevent both intentional and unintentional misuse of systems by authorized users, standards of use must be structured so that they deter intentional abuse and also instruct users so that they can avoid unintended misuse. Standards of use must provide a system that will deter and detect malicious conduct, but will also educate and instruct authorized users to reduce the chances of inadvertent abuse of the system. For example, those standards should describe the penalties associated with trafficking in access codes for the system, but they should also advise users on how to avoid inadvertently disclosing those codes. For many organizations, the standards are actually likely to be of greater value in preventing inadvertent abuse of the system than they will be at preventing intentional misuse.

Insurance

As previously discussed, a growing number of insurance carriers now provide coverage for damages and losses associated with computer system security breaches and system failures. It is important to consider this option if your organization makes significant use of computers. When evaluating insurance options, make sure that you consider coverage that will compensate for both unauthorized, deliberate abuse of your computer network by outsiders and misuse of the system by authorized users, including your employees, contractors, and other authorized users. Also make sure that any coverage you obtain compensates for both deliberate misuse and unintentional misuse. In many instances, damage caused by authorized users is inadvertent; thus insurance coverage for unintended misuse is an important element of an effective computer system insurance strategy.

Appendix 4A: Acceptable use standards

Organizations that make significant use of computers and computer networks should develop and enforce acceptable use standards. Those standards should be clearly described for all authorized system users (e.g., employees, contractors, business partners), and they should be updated regularly. The standards should be presented in written form and should be coordinated with related company policies (e.g., computer security, information privacy). Key topics to be addressed by those standards include the following.

Written standards

Acceptable use standards should be presented to users in written form. It is a good idea to require computer system users to review the written standards and to sign the document to demonstrate their understanding and acceptance of its terms. Signed versions of the document should be retained as official records of the organizations.

Training

Training sessions should be provided for all authorized users of the computer system to help them understand the acceptable use standards. Training should be provided to all new users and periodic refresher sessions should be held to help the users stay current on the standards, as those standards will likely change over time.

Personal use restrictions

The computer system standards should clearly indicate that all computer equipment and network access should be used only for company purposes and that personal use of the system is prohibited. Although there is a great deal of flexibility for each organization to make its own decisions about how rigorously it will enforce this restriction, it is generally a good idea for enterprises to adopt and express this basic limitation on use.

Summary of key legal concerns

Acceptable use standards should indicate that the system cannot be used for any activities that are illegal. This discussion should include a review of the following major legal concerns associated with unauthorized system use: (1) harassment and discrimination, (2) intellectual property misuse, (3) information privacy, (4) defamation, (5) antitrust and competition law violations, (6) export controls, and (7) financial disclosure rules. This discussion should also explain the potential legal penalties and liabilities that the organization and the individual system users may face as a result of misuse of the computer system.

Coordination with other company policies

The acceptable use standards should be coordinated with other key policies of the organization. In particular, they should be coordinated with the employment handbook which defines the various rights and obligations of employees of the organization. Some organizations may find it more convenient to make the acceptable use standards for the computer system part of the employment handbook. If an organization has employment contracts with its employees, computer use standards should be integrated into those contracts. Acceptable use standards should also be coordinated with information privacy and computer system access/security policies and practices established by the organization.

Summary of different levels of authority

Some organizations may choose to grant different levels of authority for system use to different users. For example, certain employees may have need, based on their job requirements, for access to more of the network's content than other employees. If there are different levels of system access, the standards should so indicate, and the organization may choose to summarize those different levels of access. The standards should clearly indicate that each authorized user is required to comply with the limitations on his or her usage authorization.

Notice of monitoring

If an organization chooses to monitor computer system use by authorized users, the description of acceptable use of the system should provide notice of that monitoring. It is also generally helpful to explain that the monitoring is intended to promote compliance with the acceptable use standards and thus reduce the risk of legal liability for the authorized users and the organization. If a monitoring program is implemented, the data generated by the monitoring must be managed effectively, consistent with the organization's policies and practices for important electronic records and documents.

Reporting obligations

Authorized users should be obligated to report all suspected violations of acceptable use standards. The standards should describe the process to be used to report the suspected violations. In the case of ISPs and other types of on-line service providers, particular attention must be paid to developing a clear and effective system enabling third parties to report suspected system misuse (e.g., copyright infringement, defamation, obscenity). Instruction as to how to report such suspected violations (including contact information, what type of information to be provided in the report) must be clearly presented by on-line service providers. The standards should also indicate what actions the service provider will take if the violations are confirmed, to its satisfaction.

Termination of authorization

The standards should state that all computer system authorizations can be terminated or changed by the organization at any time and for any reason. The system user should have an immediate obligation to comply with the termination or modification of usage authority. Upon termination, the former system user should have a continuing obligation to cooperate fully with the organization to complete the process of removing the former user from the system (e.g., surrendering access codes and computer equipment, returning data or other system content).

Penalties for violations

The standards should indicate that violations of any of the computer system acceptable use standards are grounds for termination of employment. They should also indicate that, in addition to termination of employment, the enterprise reserves the right to pursue any criminal or civil law remedies that might be available to it against an authorized user of the system, in the event of violation of the acceptable use standards.

Appendix 4B: Guidelines for system monitoring by employers

Computer system owners have the legal right to monitor usage of their systems. While exercising that right, however, they have an obligation to be mindful of the privacy interests of employees and all other authorized system users. When determining how to monitor computer use, system owners should consider the following guidelines.

Notice to employees

No matter what level of computer system monitoring an organization chooses to implement, it should clearly describe to its employees all of the monitoring activities. It is often helpful if the employer explains its commercial and legal reasons for conducting the monitoring. Periodic reminders of the monitoring practices should be provided to all employees, and the employees should be briefed each time that monitoring practices are modified. Employers should provide this notice to employees in written form, and should require employees to sign a statement indicating that they understand the monitoring program.

Limited monitoring

If a monitoring program is implemented, it is best if the program is as limited in scope as it can possibly be, consistent with the commercial, technical, and security objectives behind the program.

Retention of usage data

An employer must decide what data derived from the monitoring will be retained, and for how long. The employer should also identify who will have access to the data and for what purposes.

Impact of retention

When an employer retains usage data, he or she will generally be assumed to have knowledge of that data. This means, for instance, that if an employer monitors employees' e-mail messages, and if some of those messages indicate a form of illegal harassment activity, the employer will likely be deemed to have actual knowledge of that illegal conduct. With that knowledge comes a duty to report the illegal conduct and to take reasonable steps to stop it. An employer who monitors employee use of computer systems must thus continuously evaluate the data and react to it. In some instances, the knowledge that the monitoring data provides to the employer will create additional legal obligations for the employer, and failure to meet those obligations can create legal liability for the employer. System monitoring creates new legal records for the system operator. Those additional records can help an employer to enforce its rights or to reduce its potential liability if they are managed effectively. If managed ineffectively, however, those records can increase the risk of legal liability for an employer.

Appendix 4C: Tips for monitoring by ISPs and other on-line service providers

ISPs, Web-content hosts, application service providers, and other providers of on-line services have a legitimate need and a legal right to monitor use of the services they provide. As they exercise that right, however, they must also be aware of their duty to respect privacy interests of the users of their services. This duty is both a legal duty and a commercially inspired one, as service providers that respect user privacy are likely to derive commercial competitive advantage from that respect. On-line service providers seeking to balance their legitimate system monitoring demands with privacy expectations of users should consider the following guidelines.

Written notice

Service providers should describe the monitoring program in written form (on-screen statements included in terms of service are permissible, provided that they are clear and easily identified). It is helpful if the description of the monitoring program identifies the rationale behind the decision to monitor. The notice should also identify specific individuals the user can contact if he or she has questions or concerns about the monitoring.

Limited monitoring

If a usage monitoring is implemented, it should be as limited in scope as possible, given the commercial, technical, and security objectives of the service provider. The program should be structured so that it is directed toward gathering only the usage information legitimately needed for valid operational reasons. The monitoring program should be subject to continuous review and audit to ensure that it is implemented in a manner consistent with its design.

Records management

Data obtained from usage monitoring should be given the same level of security and protection that the service provider uses for its most confidential material. In some jurisdictions, privacy laws prevent the disclosure of this type of material without the prior consent of the system user; thus, the records created by the monitoring process should not be disclosed to any third party unless required by law.

Termination of service

The service provider should clearly reserve the right to terminate service or remove content from its system when its monitoring indicates that a system user is conducting illegal activities, activities that violate the terms of service or actions that threaten the commercial interests or property of the service provider.

Appendix 4D: Coordinating acceptable use standards with contractors and business partners

It is common for organizations to interconnect portions of their information systems, for business purposes. Although that interconnection can yield substantial commercial benefits, it also requires effective coordination of the acceptable use standards of the different organizations. Important issues to be dealt with during that coordination process include the following.

Written agreement

All parties who will be participating in the interconnected system should be bound by a legally enforceable contract.

Consistent acceptable use standards

The contract should require all parties to implement specific acceptable use standards for their computer networks. Ideally, the contract will define those standards. Short of that ideal approach, the parties should at least review each other's acceptable use standards to verify that the standards to be applied by all of the participants are consistent with each other and operationally sufficient. After such verification, the contract could incorporate by reference the existing standards of each of the parties, binding those parties to meet the standards that were reviewed and approved by the other participants.

Remedies for breaches

The parties should agree in advance on the procedures to be applied in the event of a breach of the usage standards. They should agree how they will respond to the breach, including remedial actions to correct the problem and what disclosures or reports they will make, and they should agree how they will share any liability resulting from the breach.

Modifications

The usage standards will likely require periodic updating and revisions. The parties should agree in advance on the procedures to be applied to develop and implement those modifications.

Termination

The parties should agree on termination procedures. They should identify what circumstances or events will trigger termination and what process will be used to unwind the integration of their computer operations. The parties should agree in advance that they will each provide all the cooperation necessary to permit successful termination of their joint arrangement, as the operational aspects of executing termination can be difficult.

5

Protecting Data

A major aspect of security in the digital marketplace is the need to protect the integrity of computer system content. Misuse of computer content can create legal liability for the owner of the computer system and for the individual computer users. Two of the most important components of electronic content protection are the effective management of information privacy and intellectual property use. Although different in some respects, information privacy and intellectual property management are also similar in that they each require proper protection of content security. This chapter examines the key legal rights and liabilities associated with information privacy and the next chapter discusses intellectual property protection.

Information privacy

The public demand for increased information privacy is currently significant and is likely to be an even greater force in the near future. Public concern on this issue has spurred attention by governments in various jurisdictions. The focus of this concern has been on information about individual people that is collected and used in a form such that the information can be linked or attributed to a specific individual. Legal requirements associated with information privacy are directed toward this personally identifiable information, not broad collections of aggregated information.

In general, the liability associated with information privacy breaches is direct. The party whose personal information has been improperly collected, distributed, or used has a cause of action against each of the parties who participated in the unauthorized activity. For example, if a Web site operator improperly collects personal information from an individual, that operator can be legally liable to the individual. Similarly, if a database owner properly collected personal information from an individual, but makes an improper use of the information or improperly transfers the information to another party, the individual will likely have a legal basis for relief from both the collector of the information and the recipient of the information in the transfer. Laws protecting the privacy of personal information thus give the owners of that information rights that can be enforced against parties who access or use that information improperly.

General privacy laws

Some jurisdictions have enacted statutes that impose requirements for the collection, distribution, and use of personal information, in general. These laws do not focus on any specific format or medium of collection for the information. Instead, they provide rules that are applicable to personal information in all formats (e.g., electronic, hard copy, verbal form). One of the most important examples of this type of broad statutory framework for privacy protection is provided by the European Union's Information Privacy Directive, which has created a framework for information privacy protection in all of the nations that are members of the EC.

The European privacy laws impose restrictions on the collection, distribution, and use of personal information. Personal information is defined as information that can be identified with a specific individual, through use of some identifier (e.g., name, street address, e-mail address, telephone number). Under the European rules, notice must be provided to individuals before the information is collected. The notice should describe what information is being collected, identify who will have access to it, and describe how it will be used. Individuals should be given the opportunity to review and correct all collected information, and they should have an opportunity to elect not to have their personal information collected (a so-called right to "opt-out" of the collection process). The rules also require that any transfer of personal information from the collector to another party can only take place if the recipient of the information is subject to legal requirements for the protection of the information that are at least as comprehensive as those applied by the EC. European privacy rules also require that all parties who collect or handle the personal data must apply reasonable security measures to protect the information from unauthorized users and from tampering.

Organizations located outside of Europe can fall within the jurisdiction of the privacy rules in two ways. If those organizations collect information directly from European citizens, then the organizations are subject to the rules. Also, if the non-European organization receives the personal information from a European organization (e.g., a U.S. corporation receives personal information about customers or employees from a European subsidiary), then both the non-European recipient of that information and the European transferor of the information are subject to the privacy rules.

European governments and the U.S. government have been in conflict over the privacy requirements. The EC takes the position that U.S. privacy laws do not provide a level of protection consistent with that provided in the European nations. For this reason, transfers of European personal information to U.S. companies or individuals are not automatically considered to be in compliance with the European requirement that transfers of information to third parties are only permissible when those transfers are governed by legally enforceable obligations upon the recipient of the information to protect the privacy of the information at a level of security equal to that required in Europe. The United States is not the only jurisdiction that has, in the eyes of the Europeans, inadequate information privacy protections. For example, the EC has determined that Australian privacy laws are also insufficient compared to the European privacy laws.

To prevent a finding that all U.S. recipients of European personal information are in violation of the European laws, the U.S. government has negotiated a *safe harbor* provision to be temporarily applied to personal data transfers to the United States. To qualify for the protection of the safe harbor, the recipient of the information must enter into contractual agreements that acknowledge that the recipient accepts the terms of the European information privacy obligations and agrees to be bound by them. These contract requirements would most commonly take the form of specific clauses that would be approved in advance by European authorities and would be required for use in contracts between European and U.S. parties. The contracts containing those clauses must be fully executed before the personal information can be distributed. The safe harbor approach thus provides a contract law basis for application of the European privacy rules to parties in the United States.

In the early stages of the safe harbor implementation process, response by U.S. companies has been far from enthusiastic. Relatively few companies have been willing to enter into the safe harbor agreements. It appears that some of the companies are waiting to see what legislative action on privacy the U.S. government will take, if any, before they commit to acceptance of the European provisions. It also seems that some of the companies do not want to be on record accepting the privacy principles at present, as they may choose to challenge the validity of some of those principles in judicial or legislative arenas in the future. Some U.S. government officials reportedly believe that the terms that the EC is considering for the safe harbor contract provisions are broader than those agreed to by U.S. officials.

Another aspect of the U.S.–European debate over the safe harbor component of the privacy rules involves the issue of whether the standard contract clauses contemplated by the safe harbor should be used by the financial services industry. The European nations currently intend to include financial services companies within the scope of the safe harbor. Some U.S. companies in that industry, however, contend that the safe harbor rules are too onerous to be implemented immediately by that industry, and they argue that the financial services industry should be subject to a different set of privacy rules that are more consistent with the requirements now being developed under the terms of the U.S. Financial Services Modernization Act.

Many companies in the United States are also concerned about the fact that the safe harbor rules would be applied to personnel data. In the United States, the basic philosophy associated with employee files has been one in which the files are the property of the employer. Application of personal privacy principles from Europe shifts the ownership of the content of those files to the employee. This difference is the basis for another significant set of criticisms raised against the safe harbor and the overall privacy regulations by many U.S. companies. The U.S. companies fear that application of the European privacy rules to personnel information would give European employees greater legal rights regarding their records than those afforded to U.S. employees of the same organization. This imbalance would likely force the U.S. employers to grant the same expanded rights to their employees in the U.S., thus necessitating a significant change in the personnel operations of the companies.

All parties who collect or handle personal information from European citizens must address the European privacy laws. At present those parties must meet the safe harbor terms or stop handling the information. Failure to meet the safe harbor provisions can result in liability to individual people in Europe and to the various European national governments. If an enterprise outside of Europe handles personal information subject to the approved safe harbor terms incorporated into contracts it executes with European organizations, those organizations will have contract law claims that can be raised if the non-European enterprise fails to fulfill its contract obligations to protect the privacy of the information.

There are some signs that the general reluctance by U.S. companies to accept the safe harbor provisions may be beginning to subside. Microsoft Corporation, for instance, has reportedly indicated that it will accept those privacy standards. Approximately 40 U.S. companies now intend to sign up to the safe harbor terms, and it is likely that the number of non-European businesses committing to those terms will continue to increase. It seems that a major reason why more companies are adopting the safe harbor provisions is the recognition that greater attention to information privacy protection is likely to be a continuing public trend, and that organizations appearing to be insensitive to the issue of privacy protection may suffer adverse commercial consequences in the marketplace.

Internet and electronic privacy laws

Some jurisdictions have opted to take a more focused approach to information privacy laws. These jurisdictions have not implemented general information privacy statutes, as the EC has done, but have instead chosen to enact laws protecting personal information obtained using the Internet or other electronic communications systems. This privacy approach is one that is based on a medium-specific strategy. Information taken in electronic form is thus treated differently than information collected using other media (e.g., hard copy, telephone). Two examples of this approach to information privacy are systems either under development or recently implemented in Canada and the United States.

Canada's information privacy legislation takes the form of the Personal Information Protection and Electronic Documents Act. This recently enacted statute focuses on collection and use of personal information in electronic form, and it is nearly identical to the European Union's information privacy directive in regard to the specific requirements associated with use of personal information. The legislation provides private parties and the government with the ability to enforce its terms through litigation against the parties involved in any collection, distribution, or use of information that violates the statutory privacy requirements.

In the United States, the Electronic Communications Privacy Act has become a popular legal basis for assertion of information privacy claims at the federal level. The act is not a broad piece of privacy legislation. Instead, it is a specialized statute directed toward protection of the privacy of electronic communications, while in transit and when stored. It is now being applied to a wide range of digital communications content, including: e-mail messages, instant messaging or chat postings, tracking software, cookies, and Web-usage monitoring systems. Under this act, both private parties and the federal government can take legal action in the event of privacy breaches associated with electronic communications content.

At the federal level in the United States, there are also several different pieces of legislation addressing various aspects of Internet and electronic privacy that are under consideration by Congress. The proposed legislation that seems to have the best chance of being enacted in the near future appears to be the Internet Privacy Act (S. 2928), sponsored by Senator John McCain. The proposed legislation addresses only personal information collected using the Internet. It adopts the notice, opt-out, and security requirements of the European rules. The act also permits parties who collect the information to use self-regulatory privacy certification programs (so-called "seal" or "certification" systems) to satisfy the privacy requirements of the legislation. Actions under the proposed law, in the event of privacy breaches, could be brought by individual parties and by the Federal Trade Commission (FTC).

Special categories of protected information

In some jurisdictions, information privacy rules are enforced with regard to specific types of personal information. These rules are content specific, meaning that they apply to certain categories of information. The United States, for example, has enacted regulations applicable to privacy for health/medical care information, financial information, and information obtained from children. These targeted forms of privacy legislation have been popular in the United States for some time.

The HIPAA in the United States applies information privacy and computer security requirements to the collection, distribution, and use of personal medical and health records. HIPAA is attempting to develop standard privacy and security practices and procedures for all personal health and medical records in the United States. Those standards will be mandatory for all parties who collect, process, store or distribute personal health or medical records in the United States. Failure to meet those standards can result in legal claims raised by the individuals who are the subject of the information or by government authorities (e.g., the U.S. Department of Health and Human Services) against the collectors, handlers, and users of the personal information.

Also in the United States, the Financial Services Modernization Act (sometimes referred to as the Gramm-Leach-Bliley Act) provided for the implementation of privacy and security guidelines to be applied to the collection, distribution, and use of personal financial information. The specific security standards to be applied to these financial records are currently being developed. These standards will apply to the collection, storage, distribution, and use of personal financial information pertaining to individual consumers. If information handlers fail to meet these standards, they can be liable to the individuals who are the subject of the information or to government authorities.

The United States has also established privacy protection requirements for information obtained on-line from children. The Children's On-line Privacy Protection Act (COPPA) empowered the FTC to establish specific rules to be applied to Web site operators and other on-line service providers who collect personal information from children. Under the COPPA rules established by the FTC, any on-line service that collects personal information from a minor must first obtain the consent of the parent or guardian of that minor. The parent or guardian must be given notice of what information will be collected, who will have access to it, and how it will be used. An opt-out option must be provided. If information handlers fail to meet these requirements, they can be liable to the individuals who are the subjects of the information and can be prosecuted by the FTC.

Contract law

Another legal basis for privacy law protection is traditional contract law. Relying on enforceable contract terms between a computer user and a service provider, some jurisdictions have applied basic contract law principles to enforce certain information privacy rights. The terms of use established by an Internet or other on-line service provider are now commonly interpreted to be an enforceable contract. Limits on use of the system or service contained in the terms of use are generally enforceable against the user, and standards and practices associated with system security and information privacy are commonly enforced against the service provider. System operators should be sure that their terms of service and other contracts with system users effectively address information privacy practices, and those operators must be mindful of the fact that they will be held legally accountable, under contract law principles, for compliance with those practices. The service provider's failure to meet the privacy standards it establishes in its terms of use will likely result in court-ordered payment of monetary compensation in an amount equal to the damage that the privacy breach caused to the system user.

These contracts also provide legal remedies for the service provider or the system operator. If the contracts clearly define the scope of user requirements with respect to protection of personal information, the system operator can enforce those contract terms against system users, if necessary. For example, if an ISP prohibits, in its terms of use, system users from disclosing personal information of other parties, the service provider could sue a system user who fails to comply with that requirement under a breach of contract legal claim in the event of an unauthorized disclosure. The service contracts thus serve to establish both the legal obligations that the service providers must meet and the rights that the service providers can enforce as to system user information privacy.

Contract law claims must establish certain facts in order to be successful. The plaintiff must show that there was a valid, legally enforceable contract. The plaintiff must also show that the defendant failed to perform an important (i.e., material) requirement identified in the contract, and that the defendant's failure to perform resulted in quantifiable harm to the plaintiff. Finally, the plaintiff must show that it acted reasonably to minimize (i.e., *mitigate*) the harm it suffered as a result of the plaintiff's failure to perform.

Consumer protection rules

Privacy rights are also enforced through the application of consumer protection regulations, in some jurisdictions. For example, in the United States, federal and state consumer protection laws have been used as the basis for some claims against parties trying to collect or use personal information. In these cases, consumer protection agencies (e.g., the FTC) take the position that any information privacy practices that are publicized by on-line service providers (e.g., in their promotional material or Web site notices) will be enforced against the service provider. The regulators assume that consumers rely on those statements and representations when they choose to make use of the service. Accordingly, if the service provider fails to apply the privacy practices it promotes, the regulators treat that failure as a deceptive trade practice that misleads and harms consumers. Under those circumstances, the regulatory agency will prosecute the service provider for misleading its customers. Service providers should, as a consequence of this oversight, make sure that they deliver on all promises they make regarding information privacy.

Tort law protection

State law in the United States provides for private tort law claims for the recovery of damages caused by misconduct or negligence by other private parties. In some states, tort law provides for remedies for certain unreasonable actions that violate the privacy of individuals. Tort law thus offers a vehicle for legal claims based on improper use of personal information. The party alleging misuse of private information can bring a legal tort claim to recover damages suffered as a result of the privacy breach. These tort law remedies for misuse of private information are, however, not available in all jurisdictions.

In these tort law claims, the plaintiff must successfully demonstrate certain facts if it is to be successful. The plaintiff must show that there was a legally enforceable duty of privacy with which the defendant was obligated to comply. The plaintiff must also show that the defendant failed to meet that duty, and that the defendant's failure directly caused quantifiable harm to the plaintiff. The defendant's failure to meet the duty of care must be the result of some form of negligent or reckless conduct by the defendant. The plaintiff must also show that it did not contribute to the harm through some form of negligent conduct and that it did not assume the risk of harm through some type of contract or waiver of rights. If the plaintiff successfully proves these facts, then the court will award monetary compensation from the defendant to the plaintiff to enable the plaintiff to recover from the damage caused by the misconduct of the defendant.

The most readily applicable aspects of tort law privacy protection for the computer system context are protection from disclosure of private facts and protection from presentation of an individual in a "false light." Unauthorized disclosure of private facts permits recovery of damages when an individual's reputation is harmed by the disclosure of information that the individual protects from public knowledge. For example, if a nonpublic figure has a medical condition but chooses to keep that information away from public disclosure, a party who releases that information to the public can be liable for damages that result from the disclosure in some states in the United States and under certain conditions. Any party participating in the distribution of that damaging personal information can be held liable under a false light tort claim. If, for example, a party released the information through an e-mail or other form of on-line messaging, both the individual who released the information and the service provider who helped to distribute it could face liability.

False light privacy tort claims involve allegations that the defendant acted in a manner that harmed the plaintiff by portraying the plaintiff publicly in a negative manner. An example of this type of action would be the distribution, by one party, of an e-mail message under the name of another person without the consent of the party named. If that misleading message resulted in harm to the reputation of the party whose name was used without authorization, a tort law claim of false light could be raised in a jurisdiction that recognizes this privacy tort.

Civil liberties

Constitutional law in the United States provides certain legal protections against unreasonable searches or seizures of property conducted by government authorities. As law enforcement and national security agencies become increasingly involved in searches of computer records and communications, constitutional law claims of unreasonable intrusions into personal privacy as part of electronic surveillance efforts by government authorities will almost certainly arise.

Constitutional law concerns in the United States have begun to surface in the context of the Carnivore surveillance system advocated by the FBI, for example. Under U.S. law, all monitoring and surveillance activities conducted by governments must be compatible with the constitutional rights of free expression and freedom from unreasonable searches and seizure of property. Communications and information in the United States have at least a basic level of legal protection from government access, to the extent that such access is found by a court to be an excessive intrusion into rights granted to U.S. citizens by the First and Fourth Amendments of the U.S. Constitution. These constitutional law claims can only be raised against government action in the United States or against private parties who are acting on behalf of governments. The extent to which constitutional law claims limit computer system monitoring by government authorities can, however, affect computer system operators and users by influencing the extent to which the system operators must cooperate with authorities when they conduct surveillance of computer use.

Privacy and location tracking

As mobile access to computer networks becomes an increasingly important aspect of computer operations, issues associated with privacy and mobile user tracking are gaining greater public visibility. Many efforts are under way to expand the ability of network operators to track the actual physical location of network users. Mobile tracking offers some significant potential value for system users and important commercial opportunities for businesses. The process also, however, raises interesting issues of privacy.

Location tracking of network users can help to more completely serve user needs. For example, tracking can help the network operator to send assistance to a network user who is in trouble, but may be unable to report the situation (e.g., a user who is unconscious due to a medical emergency or one who has been kidnapped). Mobile-tracking systems can also facilitate the distribution of location-relevant information to users at the appropriate time. For instance, as I walk through a community, a location-tracking system could provide me with information about stores or movie theaters near where I happen to be at any particular time.

Location tracking can also, however, be viewed as an invasion of privacy. If a computer system is able to monitor my location at all times, I am in effect, under continuous surveillance by the network. To many people this would be an unacceptable intrusion into their privacy. Computer systems that include some form of mobile-tracking capability thus run a risk of facing legal claims of personal privacy violations raised by users of the system who do not want to have their location monitored.

If you are considering use of a location-tracking capability as part of the mobile access component of your computer network, consider the potential personal privacy implications of such a capability before you implement it. The issue is made more complex for system operators as some jurisdictions are moving toward mandatory mobile-tracking capability, in the interest of user safety. The United States, for instance, now requires that wireless communications networks include location-tracking capability as part of their base-service offering. Although not yet fully enforced, this requirement applied by the Federal Communications Commission will lead to a situation where location monitoring will be a required network capability, and once it is fully integrated into U.S. networks, there will be increasing interest in making diverse use of the tracking information. As organizations make commercial decisions as to how the location data will be used, it is important that privacy concerns of network users be given due consideration.

Privacy practices

Computer system operators should develop and implement effective information privacy practices. Effective information privacy practices are necessary for compliance with legal obligations. Failure to develop and fully implement such practices can lead to significant legal liability for computer system operators. The following subjects are among the more important components of an effective information privacy program for computer operations.

Computer operations that collect, distribute, or use in any way personal information should implement and publicize a description of their information privacy practices and procedures. Those practices should be clearly defined in written form and should be updated regularly. All users of the computer system should be given a copy (electronic or hard copy) of the description of the privacy practices, and if possible they should be briefed on those practices. The description of the practices should include a description of the security measures used to protect the personal data.

There should be a description of the personal data to be collected. That description should include an explanation of how that information will be used and who will have access to it. It should also provide a system through which individuals can review and modify their personal data, and that system should be clearly described to all individuals participating in the system. Individuals should be apprised of how long their personal information will be retained and what will happen to it at the end of the retention period.

Individuals should be given a chance to choose not to have their personal information collected (i.e., an opt-out option). To be even more conscious in terms of privacy, some organizations apply an opt-in approach instead of the opt-out system. Under the opt-in approach, the policy is to collect no personal information unless an individual chooses to permit collection (i.e., opts-in to the collection process). An opt-out system complies with all current legal obligations for information privacy, but privacy-conscious consumers may view an opt-in system more favorably.

The privacy practices should also address transfers of personal data to third parties. If the information is to be shared with any parties other than the collector of the information, those other parties should be identified. The privacy practices of those third-party recipients of the personal information should also be described for the individuals. This disclosure should be made even if the organization only intends to share the information with groups that are affiliated with it. Also, the discussion of practices as to transfers of personal information to third parties should address the issue of transfers made as part of an acquisition or merger of the business and transfers that may take place as part of a bankruptcy proceeding.

The privacy practices should also include a process through which individuals can raise questions and complaints. Contact points and procedures through which individuals can have their questions answered and their problems resolved should be clearly established in the privacy practices. The description of these processes should include a discussion of the specific information that an individual must provide in order to raise a question or challenge an action.

Clear penalties for failure to comply with privacy practices should be established and described. Failure to comply with the standards should be ground for termination of employment or termination of a service contract if the party at fault is a contractor. The organization should reserve the right to pursue other legal remedies (e.g., private law suits or criminal prosecution) against a party who fails to comply with the privacy practices.

Periodic audits should be a part of any organization's privacy practices. These audits should review the way the organization actually implements its privacy practices, and the audits should be performed on a regular schedule. For maximum effectiveness, it is best to have these privacy audits performed by neutral outside parties (e.g., business consultants, accountants). The results of the audits should be used as the basis for adjustments to the privacy practices.

The organizations should consider participating in one of the privacy certification or seal programs operated by certain industry groups and trade associations (e.g., the eTrust program). Participation in these programs can help to improve the quality of the organization's privacy practices, and it can give customers and business partners of the organizations a higher degree of confidence in the privacy measures. An additional potential benefit of participation in these programs is that the effort can help to reduce the scope of legal liability if there are any privacy breaches in the future. Certification program participation can be used as evidence of good faith efforts to protect personal information privacy and to comply with all legal and regulatory requirements associated with that information.

Appendix 5A: Privacy practices and procedures for personal information

All organizations that collect, distribute or use personal data about individuals should establish written privacy practices and procedures. Those practices and procedures should, at a minimum, address the following topics.

Notice

Clear, written notice should be provided to all individuals from whom personal information is being collected. The notice should discuss the following subjects: (1) description of the information to be collected, (2) identification of all parties who will have access to the information, (3) description of what the information will be used for, (4) discussion of how long the information will be retained, (5) explanation of how the individual can elect not to provide the information, and (6) description of the process through which the individual can obtain access to the information for review (or modification) purposes.

Specialized requirements

Laws in various jurisdictions require special privacy treatment for specific types of personal information (e.g., health/medical, financial, children's information). The privacy practices of your organization must comply with the specific rules regarding these special categories of personal information, to the extent that your organization handles such information.

Choice

The individual should have the right to choose not to provide the personal information that has been requested. This is the right to opt-out of participation. Some organizations may choose to go even further, adopting an opt-in approach. With an opt-in structure, no personal information is collected unless the individual first chooses to offer the information. With an opt-in approach, the default position is one in which no personal information is collected.

Security

Appropriate security measures should be implemented to protect the privacy and integrity of all personal information that is collected. Descriptions of the security measures should be available for review by individuals, upon request.

Transfer controls

All transfers of personal information to parties that are unaffiliated with the organization that collected the information should be prohibited unless prior approval from the subject of the information is obtained.

Access by subject

The individual who is the subject of the personal information should have the right to review the information. This right of review should include the right to modify, correct, or delete the information.

Complaints

Policies and practices for the management of personal information should include a process through which individuals can review the information that has been collected about them and can register complaints regarding the information or the collection and use process. This process should be clearly explained to individuals and should be consistently implemented.

Auditing

The information collection and management practices should be monitored on a continuing basis. Periodic audits of the system should be conducted on a regular basis to ensure that the procedures are actually being applied properly. The audits should be conducted by an outside party that is independent of the organization.

Certification

Organizations should seriously consider participation in one of the privacy certification or seal programs that are now sponsored by industry organizations, trade groups, and consumer organizations. They perform monitoring, review, and audit functions in support of information privacy activities of businesses. Enterprises with information privacy practices that comply with the certification requirements of these programs are authorized to identify themselves to the public with the certification or seal of the program.

6

Protecting Intellectual Property: Digital Rights Management

In addition to personal information, a second key category of computer system content is intellectual property. Much computer content is governed by intellectual property law principles: copyright, trademark, patent, and trade secrets. Effective intellectual property management practices are essential in order to reduce the risk of legal liability associated with improper use of such system content. System owners and system users who make improper use of the intellectual property of other parties can face significant legal liability. Operators of computer networks must also manage those networks effectively to ensure that they preserve all of the legal rights available to protect the intellectual property that they own. This chapter focuses on the basic legal rights and responsibilities associated with security for computer system intellectual property.

Managing licensed products

Most computer systems store, distribute, or use some amount of intellectual property that has been developed by another party. To use that content lawfully, the user must obtain permission from the owner of the intellectual property rights for the material. The contract between the copyright owner and the authorized user is commonly identified as a license, and it provides the legal foundation for all authorized use of intellectual property. Computer systems must be effectively managed to ensure that all licensed property stored, distributed, or used within those systems is handled in a manner consistent with the relevant licenses. Failure to manage the use of licensed intellectual property properly can lead to legal liability for the individual users of the computer system and for the owner of the system.

Intellectual property licenses tend to focus on a few key use factors. Licensed property must be managed so that only the authorized number of users has access to the material. The property must be used only for authorized purposes. Licenses commonly restrict duplication, distribution, and modification of the licensed material, specifically limiting the number of copies of the material that can be created and generally prohibiting alteration of the original material. Finally, the material can only be used for the authorized period of time, or term. These key elements of an intellectual property license must be effectively cataloged and enforced by all licensees. This process of managing license compliance is necessary to reduce the risk of copyright infringement liability.

Work-for-hire

Some computer system owners make use of software and other forms of electronic content that they develop on their own, making use of their employees or contractors to create that material. In most cases, the employer intends to assert ownership rights over the intellectual property created by its employees within the scope of the employee's job requirements. This principle is known as *work-for-hire*. It is based on the notion that the employee has been hired to perform a variety of duties for the employer, and that any property created by the employee as part of his or her employment is to be owned by the employer.

If a work-for-hire relationship is not effectively established, the employee may own the intellectual property he or she creates. In that situation, employers would be forced to negotiate a license with the employee if they want to have access to the property. Use of the property by the employer without such a license would be unlawful and could subject the employer to monetary damages or court orders to cease use of the property.

To create a work-for-hire relationship, the employer must make sure that it treats the individual as an employee (e.g., provides a salary, provides equipment and operational support, pays benefits, withholds tax payments). The employer must also have an employment contract, employee manual or handbook, and any other documentation that establishes or describes the job duties or scope of employment of the employee, clearly indicating that the relationship is an employment relationship, that the party involved is an employee, that all intellectual property created by the employee is work-for-hire and is owned by the employer, and that the employee assigns to the employer all rights that he or she may have in the works created.

If an organization makes use of contractors, material created by those contractors is generally not considered to be work-for-hire. Original work developed by contractors is generally the property of the contractor. If the parties want the ownership of the copyright for the material created by the contractor to be held by the client organization, they must take special steps in advance of the creation of the material to accomplish that goal. There should be a written contract that indicates that the parties agree that the work created by the contractor will be owned by the client. In addition, the contract should include an assignment and transfer to the client of all ownership rights that the contractor may have in the work.

If the intellectual property developed is subject to patent protection, a common approach is to identify the employee or the contractor as the inventor, but to execute an assignment of ownership of the patent to the employer/client. This arrangement should be agreed upon by contract prior to commencement of the invention development process. When patent applications are filed, the applications should reflect this arrangement by properly identifying the parties who are the inventors and those who are the assignees of ownership rights for the patents.

Copyright piracy

When copyrighted material is duplicated and redistributed without the prior consent of the owner of the copyright, we generally consider the unauthorized copies of the material to be bootleg or pirated copies. If copyrighted material is pirated, the individual who made the unauthorized copies is liable to the copyright owner, and the owner of the computer system used to duplicate, distribute, or make other use of the pirated material may also be liable to the copyright owner.

Copyright owners can bring private legal actions for copyright infringement in all countries that are signatories of the WIPO Treaty. Remedies available to the copyright owner for piracy include court orders to the pirate to cease distributing the pirated material and to destroy all remaining copies of the material. The remedies available also include monetary awards to compensate for damages caused to the copyright owner as a result of the piracy (e.g., compensation for lost sales resulting from the distribution of the pirated versions of the product). Some jurisdictions also permit courts to award punitive damages to punish the pirate, and prison time can also be ordered by courts in some jurisdictions.

Regulating copyright circumvention technologies

An interesting and troubling public policy approach is being explored in some jurisdictions to attempt to deal with the issue of pirating copyrighted material. This approach involves regulation of technologies that can be used to circumvent copyright-ownership systems and technologies applied by copyright owners to their material. Some jurisdictions have gone so far as to make the manufacture, distribution, or use of a technology that can circumvent copyright-ownership protections illegal. Other jurisdictions are experimenting with the use of fees or taxes assessed against the manufacturers of those technologies.

The Digital Millennium Copyright Act in the United States, for instance, makes the manufacture, distribution, or use of technologies that have copyright circumvention capability illegal. The precise scope of this prohibition remains, at present uncertain, but it appears to apply both to hardware and software that can be used to defeat copyright protective measures, decryption systems, for example. This prohibition was raised, but has not been resolved in the *Napster* litigation associated with on-line music distribution (*R.I.A.A. v. Napster*). The allegation that the Napster peer-to-peer sharing software and on-line system constituted a technology that assisted in the circumvention of music copyright protection was raised by the music industry in that case.

Another current U.S. case (*Universal City Studios, Inc. v. Reimerdes*) involving litigation over the DeCSS software used for decryption of DVD content (and the on-line system applied to distribute that software) has also raised this circumvention argument, but the case has not yet been resolved. The anticircumvention provision has also been raised in U.S. litigation, which has not, at the time of this writing, been resolved, challenging the Streambox, Inc. "Ripper" program that converts the multimedia files of RealNetworks into other formats (*RealNetworks, Inc. v. Streambox, Inc.*).

The current version of the draft Convention on Cyber-Crime of the EC also prohibits technologies with copyright circumvention capability. The convention makes copyright infringement a cyber-crime (Title 2, Article 10), and it makes production, sale, distribution, or possession of any device or equipment, with the intent to commit an act in violation of any of the terms of the convention, an offense (Title 1, Article 6). This provision is one of the more controversial aspects of the current draft of the convention.

Germany now applies a fee assessed against manufacturers of equipment that can be used to circumvent copyright-protection measures. The fee is charged on a per-unit-offered-for-sale basis. Revenues from those fees are collected by German copyright societies and are put into a fund that is to be distributed to copyright owners. This charge is, presumably, passed through to all consumers of the equipment. In this way, the fee effectively functions as a tax that is borne by all purchasers of the equipment in question, even by those individuals who do not use the devices to pirate copyright-protected material.

One of the German copyright societies, VG Wort, has reportedly initiated legal action in Germany to force computer manufacturers to pay a per unit royalty of 30 euros for each personal computer sold in Germany. The VG Wort legal action initially targets the computer manufacturer, Fujitsu Siemens Computers. It appears to be an extension of a previous action against manufacturers of compact-disk burners (including Hewlett-Packard Company) that resulted in an arbitration ruling requiring the burner manufacturers to pay a per-unit-sold fee of 6 euros (the arbitration ruling has reportedly been appealed). This extension of the fee into the computer manufacturing marketplace represents a significant expansion of the equipment royalty approach.

Many industry observers are troubled by sanctions applied to technologies that have the ability to circumvent copyright protections. In most instances, these technologies have functions in addition to the circumvention functions. Sanctions applied to those technologies can create disincentives to future technological innovation and barriers to widespread adoption of new information technologies. For many industry participants, this is a bad result. Banning or taxing specific technologies often provides disincentives for future innovation, and that result can cause significant harm to future economic development. Prudence suggests that policymakers should be wary of these technology-specific sanctions, and should work hard to develop other methods to protect legitimate interests of copyright owners that have fewer adverse consequences for continuing technological innovation.

Open-source content

Open-source software is generally distributed using a license with no fee. The source code for open-source software is commonly made available to all developers who agree to make the source code for their modified versions of the product available to others at no charge, as well. The dramatic rise in popularity of the many different open-source software products makes management of open-source content an important intellectual property challenge.

When dealing with open-source software, the user must make sure that it complies with the open-source requirements, with regard to the material it develops, based on the original open-source product. Even with open-source distribution, the software remains intellectual property that is distributed subject to a license. The user continues to have an obligation to comply with the terms of that license. Although there may be no license fee associated with its use, open-source software is licensed material and failure to comply with the terms of the license can result in copyright infringement liability.

For a software developer considering making its software available on an open-source basis, an important issue associated with the open-source distribution model is that of potential fragmentation of the original product. Fragmentation is the process through which different forms of the original product are created as a result of the many different developers who work with it. A certain amount of product fragmentation is to be expected with open-source distribution, and that basic level of diversity is a major strength of the open-source approach. If, however, the fragmentation becomes too great, resulting in different "flavors" of the original product that no longer work smoothly together, the value of the original product may be reduced, and the benefits of using the open-source system will not be fully realized. Some suggest that the history of the development of Unix provides an illustration of how fragmentation can occur.

Another difficult aspect of managing open-source material is the need to ensure that the open-source material does not become comingled with material that is being treated as proprietary. If an enterprise inadvertently integrates code that it intended to treat as proprietary material into an open-source product, the enterprise will lose its exclusive proprietary rights to that material, as the proprietary material will be subsumed into the open-source product. Similarly, if an organization accidentally incorporates proprietary code from another party that it has licensed for nonopen-source purposes into an open-source product, the organization will be liable for a violation of the software license. Effective separation of open-source and nonopen-source material is an essential element of intellectual property management in an open-source environment. That separation is likely to become a greater challenge as open-source use increases.

Peer-to-peer content sharing

An important current intellectual property management issue relevant to computer security is the rise in popularity of distributed computing, shared-content models. This peer-to-peer structure provides computer users with greater control over the content that they share with other network users. Individual network users can apply these content-sharing systems to easily access content stored on the computers of other members of the network, without always relying on an intermediary party or device to facilitate the exchange. Content stored on one computer in the network thus becomes more readily accessible to all other users of the network under these peer-to-peer systems. These distributed-computing systems push greater control to the network users, reducing the role of servers, and making the data sharing less hierarchical than it is in the standard Web-distribution configuration.

The peer-to-peer systems can also make it more difficult to ensure effective compliance with copyright license terms. As network end users gain greater control over content sharing and require less contact with network intermediaries (e.g., Web servers), the distribution process for licensed material becomes more diffuse and less readily controlled by any middle party. Experience with on-line music distribution using the Napster system illustrates this point. The rapid development and acceptance of peer-to-peer systems poses one of the greatest current challenges to intellectual property management for digital content. In the world of distributed computing, both individual users and operators of the distributed networks can bear liability for misuse of intellectual property that is shared over the network.

In the distributed-computing setting supported by peer-to-peer systems, network end users bear greater responsibility for intellectual property license compliance. One of the current challenges is the development of a system for content-usage monitoring and payment collection that provides content providers with greater confidence that the license terms associated with the distribution of their property in this setting will be enforced. Content providers must develop methods to facilitate management of their content in this type of environment, and end users must learn to respect the ownership interests of the content providers. Most likely, a combination of modified business models and new technologies will provide the level of content management demanded by content providers. For example, some form of subscription payment system, where end users pay a periodic flat fee for the right to access content provided through a distributed computing system seems to offer promising prospects. Also possible is some form of negotiated *blanket* license or *compulsory* license that sets specific terms of use and royalty-compensation structure for entire classes of shared content.

From an intellectual property management perspective, the distributed-computing model takes us back to some of the earlier days in computer networking. With this type of distributed system, licensing models that calculate royalty payments based upon the number of potential users of the content seem to be the most appropriate. Licensing based on the number of content users is a model that is well established in the world of software licensing, and will likely play an important role as peer-to-peer content use expands more widely. It seems clear that the wider content sharing made possible by distributed-computing systems can be effectively managed if licensing practices are properly structured to accommodate that sharing.

Linking, framing, and cached content

Various standard and common forms of Web activity carry certain legal liability concerns. These concerns can generally be managed effectively through prudent Web operations. For example, fundamental Web tools such as links, content frames, and content caching raise legal issues, under certain circumstances. Web content providers and users must be mindful of those issues.

Use of hypertext links between Web pages is among the most fundamental aspects of Web structure. Under certain instances, however, there may be legal issues associated with Web page linking. For example, one of the issues raised, at times, is the extent to which a link from your Web site to a site owned by another party can cause you to be liable for legal violations (e.g., defamation, copyright infringement) that occur at the other site. In the case of alleged copyright infringement, in which a defendant has posted allegedly infringing material at its Web site and the court has ordered the defendant to take the material off the site, a U.S. court concluded (*Intellectual Reserve, Inc. v. Utah Lighthouse Ministries, Inc.*) that the defendant may also be liable (as a contributory infringer) if other parties took the material from the defendant's Web site and made that material available on their own sites. Courts in Japan and Holland (http://www.nikkeibp.asiabiztech.com/99001 and http://www.news.com/News/Item/0,4,0-37622,00.html?st.ne.lh..ni) seem to have taken a similar approach, finding liability based on Web site links.

The number of cases that have addressed the issue of liability based on links remains extremely limited, so it is inappropriate to conclude that such liability will exist in many different circumstances. Instead, the best approach is prudent monitoring of the content of Web sites to which your site links and use of disclaimers at your site that indicate that you are not taking any responsibility for the content of sites to which your site links. Based on regular monitoring of those other sites, you should terminate links to sites that you believe contain content that is either illegal or inappropriate.

Web site links have also raised questions of legal liability based on claims of unfair trade practices. In the United States, a dispute between Microsoft Corporation and Ticketmaster Corporation and another one between Tickets.com and Ticketmaster highlighted this legal theory. In these cases, the owner of a Web page objected to links established by another Web content provider that connected pages from another Web site. The owner of the site that was the target of the link claimed that linking without prior permission was an unfair business practice. Although these cases were settled and thus no court has formally established the boundaries of permissible Web linking, certain general principles appear to be commonly accepted.

One of those principles appears to be that links established to the home page of another site are always permissible. Also, to the extent that a Web site operator posts linking instructions, parties who wish to link to that site should comply with those instructions, to the extent that they are reasonable. The greatest risk of legal liability resulting from Web links seems to arise when the link is made deep into the other site (i.e., to a page other than the home page). Before creating deep links, the party establishing the link should review and try to comply with all notices or linking instructions posted at the target site, and the party should consider the circumstances. For example, if the target site is that of a business competitor or rival, the link may not be a particularly good idea. Similarly, if the target site relies heavily on advertising revenue, the owner of the site may be more defensive regarding links, to the extent that the links reduce page hits or bypass page views and thus reduce advertising revenues generated by the site.

A patent dispute has also arisen with regard to Web-linking technology (*British Telecom v. Prodigy*). British Telecom alleges that its U.S. Patent Number 4,873,662 covers hypertext linking and use of those links for Web content browsing. The dispute is in litigation and if the patent is found to be valid and applicable as British Telecom alleges, obviously it will have a significant impact on use of linking by all parties. Broad application of such patent could result in potential patent infringement liability for all parties who establish Web links.

Legal disputes have also been associated with use of frames and other methods of altering the on-screen appearance of Web content. The most common legal theories applied to these disputes have involved copyright infringement and unfair trade practices. These claims would be raised by the owner of the framed content, and they would be raised against the party conducting the framing. Under a copyright infringement claim, the owner of the material that is framed argues that the use of the frame creates a derivative work that can only be created with the permission of the copyright owner. The unfair trade practice claim contends that the frame alters the visual representation of the content that is framed, and that the alteration confuses customers or harms the commercial interest of the owner of the framed content in some other way (e.g., it obscures advertising presented on the framed site, and that act reduces the advertising revenue received by the owner of the framed site).

The leading case on the issue of framing was based on a dispute between TotalNews, Inc. and several major U.S. news publications (*Washington Post Company v. TotalNews, Inc.*). That dispute was ultimately settled, so we do not have a clear court ruling on the legal limits of framing. For now, perhaps the best approach is for Web site developers who do not want their content framed by others to make use of the reportedly available technical measures that can be used by a site operator to block frames imposed by another Web site. Parties considering use of frames should consider the content that they are planning to frame. If the frame might mislead or confuse users of the targeted content, it should not be applied. Also, if the frame could result in commercial harm to the owner of the targeted content (e.g., adversely affect its advertising revenue), the frame should not be applied.

Caching of Web content has also created some controversy. Some owners of copyrighted material expressed concern that caching can result in distribution of outdated (stale) versions of their on-line content. System operators and end users generally respond that caching helps to improve the efficiency of the Web content delivery process, helping to reduce the costs of system operators and improving the quality of system performance for end users.

The process of caching involves the creation of copies of Web content. This action is ordinarily within the scope of the copyright ownership rights of the content owner. A strict application of copyright law could support a copyright infringement claim by the content owner against the party who creates the cached copies. As a practical matter, however, it does not appear that any formal legal challenges to caching have yet been raised, although numerous parties have expressed concern about the process. It is likely that if a formal legal challenge is raised in the future, a key issue to be addressed based on the facts of the specific dispute will involve a determination of whether the caching in question constituted *fair use* of the content. If it were found to be fair use, the unauthorized duplication of the material would be permissible. A fair use argument, in the caching context, seems to be a strong argument as the primary purpose for caching is to improve the performance of the content-distribution system.

Domain name management

Domain name management has become a major intellectual property issue for computer operators. The primary legal principles associated with domain names involve trademark law and anticybersquatting regulations. Parties who misuse the trademarks of others in their domain names can face legal liability to the trademark owner under both trademark law and anticybersquatting theories.

Domain names commonly instigate conflicts between legitimate trademark owners who are trying to make simultaneous use of the same mark in their Internet operations. Trademark law permits more than one party to use identical or similar marks, provided that their use does not result in confusion to consumers. Companies that are in different industries (i.e., different classes of use) or do business in different geographic regions can simultaneously use the same mark without any liability to each other. We find, however, that as more businesses move more of their business operations on-line, the prospects of consumer confusion increase dramatically. The on-line environment thus makes it more difficult for parties who have legitimate trademark rights for the same mark to use that mark without causing consumer confusion. The race to incorporate trademarks into domain names provides a clear example of the on-line trademark problem.

Domain names are generally registered on a first-come-first-served basis. If two businesses have a legitimate right to a particular trademark, and if both of those businesses want to use the mark as a domain name, the one that registers it first will prevail. Unfortunately, under those facts, the other business will not be able to use the mark as its domain name, even though it too has established legal right to the mark. In this setting, the second owner of the mark is out of luck, and must find another domain name, although it continues to have the right to use the mark in its geographic region and for its class of service. In an effort to try to make it possible for more legitimate trademark owners to share use of their marks in domain names, the Internet Corporation for Assigned Names and Numbers (ICANN) is creating additional top-level domains. This process would enable more than one trademark owner to use the same mark in its domain name, each use being within a different top-level domain. Even this expansion will not, however, eliminate the problem of accommodating multiple legitimate trademark owners in the domain name registration process.

Domain name conflicts also exist when a party with no legitimate trademark claim attempts to use the trademark of another party, without authorization, in a domain name as part of an effort to derive an unfair economic gain. This process has been commonly described as *cybersquatting*. The United States has enacted a national law that prohibits cybersquatting (the Anticybersquatting Consumer Protection Act). The act permits an owner of a trademark (or other "famous" mark) to recover monetary damages from a cybersquatter. The law also enables the owner of the mark to force the cybersquatter to surrender registration of the domain name to the owner of the mark. Claims under the act must be brought in the U.S. federal courts, and the number of cases based on the act has increased substantially over time (*Cello Holdings v. Lawrence-Dahl*; *Network Solutions, Inc. v. Virtual Works, Inc.*; *Northland Ins. Cos. v. Patrick Blaylock*; *Caesar's World, Inc. v. Caesar's Palace.com*).

Cybersquatting is also prohibited under domain name registration rules adopted by ICANN. Those rules are enforced by various arbitration groups. One of the most popular of those domain name dispute resolution systems is the arbitration system operated by the United Nation's WIPO. Mark owners seeking to recover a domain name built on one of their marks can use the WIPO domain name arbitration process to raise their claim under the ICANN rules (http://arbiter.wipo.int/domains and http://www.icann.org/udrp/udrp-policy-oct24oct99.htm). Arbitrators selected by WIPO are available to resolve these disputes using a process that is quick and relatively inexpensive. The WIPO arbitration process for domain name disputes has addressed a number of highly visible cases (*Harrods Ltd. v. Robert Boyd; World Wrestling Federation v. Michael Busman; Julia Roberts v. Russell Boyd*).

Trademark owners can also assert their rights to domain names by using trademark law claims. In all nations that are signatories of the WIPO Treaty, a trademark owner can sue for trademark misappropriation to get the court to order the defendant to stop misusing the mark and to recover monetary compensation for damages caused by the misuse. A cybersquatter can thus be challenged by a mark owner through traditional trademark litigation. The claim is based on an argument that the cybersquatter's use of the mark in a domain name results in confusion to the consumer. In addition to trademark misappropriation or infringement claims, a mark owner in the United States can also raise trademark dilution claims. The U.S. claim for dilution does not require a factual demonstration that there was customer confusion. Instead, a dilution claim can be supported if the mark owner can show that the defendant's use of the mark resulted in harm to the economic value of the trademark (*Mattel, Inc. v. Internet Dimensions, Inc.*). Trademark law claims based on domain name disputes are also increasingly common outside of the United States (*Guy Laroche v. G.L. Bulletine Board; Amelie Mauresmo v. InterNIC and Jacob Medivi*; and *Procter & Gamble v. Shanghai Chenxuan Zhineng Science & Technology Development Company*).

Trademark owners should monitor on-line use of marks that are similar to their trademarks as part of their standard trademark-protection activities. Effective protection of trademark rights requires continuous surveillance to ensure that potential violations are identified quickly. Prompt enforcement actions should be taken immediately upon discovery of any inappropriate usage, and those enforcement actions are essential in order to preserve trademark rights.

Metatags, keywords, and Web search systems

The on-line search process has also raised issues of legal liability in several different contexts. One such context involves the use of trademarks in Web site metatags. Metatags are commonly used by Web site developers to make it easier for individual Web users to find their sites. Search engines access the keywords embedded in site metatags, and use those metatag keywords during the retrieval process associated with the keyword searches conducted by individual Web users. Site developers try to embed metatags that are likely to be used as keywords by the type of searchers whom the developer wants to attract to the site.

Legal problems arise when a Web site developer embeds keywords in the site that are also trademarks of other parties. Through this process, if an end user conducts a search using the trademark as a keyword, the site containing the embedded metatags will generally be included in the search results, along with the site of the trademark owner. There is a potential trademark law problem if your use of the trademark as a metatag confuses or misleads consumers.

In the United States, there have been several legal cases addressing this issue. In instances when the trademark is distinctive and has no *generic* meaning (i.e., conventional meaning not associated with the company or the products that use it as a trademark), use of the mark as a metatag by another party is likely to be a trademark law violation (*Playboy Enterprises, Inc. v. Calvin Designer Label; Insituform Technologies, Inc. v. National Envirotech Group, LLC; Oppedahl & Larson v. Advanced Concepts*). However, when the mark in question has a generic meaning and the metatag use is not likely to confuse the general public, use of the trademark as a metatag may not pose a trademark law problem (*Playboy Enterprises, Inc. v. Netscape Communications Corporation*). Thus, a mark such as "playboy" is both a protected trademark and a word that has a generic meaning that is distinct from the trademark. Use of this type of mark in a metatag is unlawful if that use confuses consumers.

The best approach to reduce the risk of liability is to avoid using trademarks of other parties as metatags in your Web site. A bit of creativity as to vocabulary and word selection should make this safeguard easy to implement. In those instances when a word is both a trademark and has a generic meaning, it may be reasonable to use the word as a metatag, provided that the generic meaning suits the content of your site and that your business is such that use of the word in your site's metatags will not lead to confusion on the part of the general public. The one way to ensure that you will face no liability, however, is to adopt a strategy of never using the trademark of another party as a metatag in your Web site.

Trademarks have also become an issue in the context of advertising purchases associated with keywords in the search engine setting. Some Web search engine operators sell keywords for advertising purposes. This means that an advertiser can buy a keyword, and that purchase places advertisements for the buyer on the screen along with the results generated by the search run on that keyword. Thus for example, if I am a golf club manufacturer, I can buy the keyword "golf," and when an individual conducts a search using the word "golf," my advertisements will appear on the screen along with the search results.

Disputes arose when advertisers began buying keywords that were the trademarks of other businesses. The leading cases on this issue involve disputes between the Estée Lauder company and a distributor, the Fragrance Counter, as well as one involving Playboy Enterprises, Inc. and Netscape Communications Corporation. These cases seem to indicate that an advertiser who purchases on-line advertising rights for keywords that are trademarks of another party can be liable for trademark misappropriation or dilution. However, if the trademark in question has a generic meaning in addition to its trademark function (e.g., "playboy" is both a trademark and a common word with a generic meaning) and if the advertising purchase is not likely to result in consumer confusion as to the products of the company that owns the trademark, then there is no trademark law liability.

There are several basic lessons to be derived from these cases. If you are considering planning a keyword advertising purchase, avoid keywords that are trademarks of other parties. If you are considering a keyword that is a trademark of another party, but that mark also has a generic meaning and you are confident that the purchase will not confuse consumers, then you may make the purchase, but recognize that there is a significant risk that you may be challenged and forced to prove your case in court. If you are a trademark owner, conduct periodic on-line searches using your various trademarks and monitor the advertising that appears in association with those search results. If the searches generate results that you believe may be confusing or misleading to consumers looking for your products, you should investigate the situation in greater detail to try to determine if there has been an inappropriate keyword advertising purchase by another party.

Property rights claims

Some jurisdictions are now beginning to consider property rights claims in association with management of property that may or may not qualify as traditional intellectual property. In the United States, for example, courts are beginning to see claims that allege that material that could qualify for protection under copyright, trademark, or other type of intellectual property law theories should also be protected under personal property law principles. Similar arguments are now being made to assert that property that would not ordinarily qualify for intellectual property law protection (e.g., collections of data) should be protected by personal property law theories. These issues are currently far from being resolved, but computer system owners and users should note them.

Many jurisdictions provide owners of property the legally enforceable right to control access to, and use of, their property. Unauthorized users of the property are subject to legal penalties. The property law concepts of unauthorized use of property are often described as the principles of *conversion* if the property at issue is personal (i.e., moveable) property and *trespass* if the property in question is land (i.e., real estate). The penalties commonly associated with these violations of property-ownership rights include court orders to cease use of the property and monetary damage awards to be paid to the owner of the property as compensation for the unauthorized use. In the United States, some courts have indicated a willingness to consider Internet domain names as personal property subject to traditional property laws (*Network Solutions, Inc. v. Umbro International, Inc.; Dorer v. Arrel*). Other courts are now considering whether the data files that contain Web site content (e.g., *eBay v. Bidder's Edge*) or streaming media content (e.g., *House of Blues v. Streambox*) qualify as personal property subject to basic property law protection.

Users and developers of digital content should be mindful of the fact that courts may begin to enforce property rights as to digital content (e.g., data files, databases) that extend beyond traditional intellectual property law rights. For users of that content, this legal trend raises significant potential legal liability. It means that each time a user (or the user's software agent) accesses or duplicates the property, he or she must first consider the ownership interest of the content developer and try to comply with any restrictions on use established by that developer. Failure to comply with restrictions on use applied by the property owner could result in the user being found to be liable to the property owner based on a theory of conversion or digital trespass.

For developers of this content, the legal trend toward application of property rights creates new rights that can be enforced by the developer. If a user fails to comply with the reasonable restrictions on use of the digital property established by the property owner, the user could be legally liable to the owner for damages. Consideration of this property law theory by courts is only in its very early stages. The fact that courts are entertaining this issue, however, means that digital content developers and users alike must be aware of the issue and should begin to plan to protect their interests if the courts extend the current trend.

Business method patents

The U.S. Patent and Trademark Office has issued a significant number of patents for business methods executed in the electronic commerce environment. These patents assert proprietary control over specific methods of conducting business activities and operations (e.g., payment systems, accounting processes). Major e-commerce enterprises including Amazon.com and Priceline.com have received some of these patents, as have many less visible companies. Patents for "one-click" payment, on-line reverse auctions, and on-line affiliate payment processes are among the most visible of the patents for e-commerce business methods. These patents have caused significant controversy in the e-commerce industry and in the patent law community. Their existence also raises important issues of potential patent law liability for e-commerce operations.

Some of the patents granted to date cover fundamental and widely used e-commerce activities and operations. This condition suggests that many e-commerce operations are at risk of U.S. patent law liability. The validity of many of the e-commerce business methods patents is currently being challenged (*Amazon.com v. Barnesandnoble.com*) in U.S. courts. If the existing business method patents are strictly enforced and if more business method patents are issued in the future, e-commerce operations will be significantly affected. If the patents are enforced, e-commerce operations that use systems covered by the patents will be required to pay some form of royalty to the patent owners or face liability for patent infringement.

So far, only the United States has moved forward aggressively granting patents for business methods in the electronic commerce setting. If other countries follow this lead, the development of e-commerce could be adversely affected. It now appears, however, that the number of business method patents being issued in the United States is decreasing. It is too early to know if this represents a shift in policy, but it could be an indication that the e-commerce method patents may be reevaluated. Court action testing the validity of the patents already issued in the United States will provide a clearer indication of the extent to which business method patents pose a longer-term threat to e-commerce expansion. For the present, parties involved in e-commerce activities should monitor the evolving legal status of business method patents, paying particular attention to the extent to which courts choose to enforce the patents that have already been granted. If those patents are vigorously enforced, e-commerce businesses must act to negotiate patent licenses for systems that they intend to continue to use, and those businesses may want to pursue a more active effort to obtain their own patents in this field.

Trade secrets

Trade secrets are pieces of information that provide commercial competitive advantage to their owner by virtue of the fact that the owner possesses them and its competitors do not. They are secrets that give their owner a commercial edge, and they can be in virtually any form (e.g., customer list, formula, manufacturing process, design). Law in many different countries prohibits unauthorized disclosure of trade secrets. Parties authorized to access and use trade secrets generally do so subject to contracts (e.g., confidentiality or nondisclosure agreements). To the extent that your computer operations handle trade secrets of your organization or those of other parties, the use of those secrets must be effectively managed to reduce the risk of legal liability.

The practices applied to manage trade secrets processed by your computer system are very similar to those applied to the management of other forms of intellectual property. Specific security and use procedures should be established to protect the secrets from disclosure. All uses of the secrets should be conducted in a manner consistent with the contractual obligations that your organization has accepted in the agreements that granted access to the material. Failure by your staff to comply with the protective measures should be made grounds for termination of employment. Compliance with trade secrets obligations should be monitored continuously and periodic audits of the security and management systems should be conducted on a regular basis to help protect the integrity of the process.

Failure to protect trade secrets effectively can have severe legal consequences. If you fail to protect your secrets effectively, those secrets may become public knowledge and your legal right to stop the disclosure or to recover compensation for damages caused by the disclosure may be lost. If you fail to protect the trade secrets of another party you may be liable to that party for breach of contract or business tort damages. You may also be subject to fines or even incarceration based upon penalties imposed by statutes such as the Economic Espionage Act in the United States. Always remember that trade secrets law requires the effective protection of both your secrets and those of other parties. You have obvious incentives to protect the secrets you own, but sometimes there is a tendency to be less diligent as to the secrets of other parties. When you possess sensitive proprietary material of other parties, it is generally best to provide the same level of security for that material that you provide for your own secrets.

Appendix 6A: Managing copyrighted material

Computer system content protected by copyrights must be managed effectively to ensure that the material that is owned by the system owner is effectively protected and to ensure that the material that is licensed from other parties is used in a manner consistent with the terms of the license. The following actions are essential to effective management of copyrighted material.

Protect original works

For copyrighted works created by your organization, you should: (1) preserve an original version of the work, (2) register the work with the U.S. Copyright Office, (3) use appropriate copyright notices when the work is distributed, and (4) aggressively enforce copyright through use of licenses and litigation when necessary.

Protect works-for-hire

Copyrighted material created by employees or contractors must be effectively managed if your organization intends to assert ownership of that material. Key steps in this process include the following.

Clear contract terms

All contracts with employees and contractors should specifically identify the work to be created as work-for-hire. All employee manuals, handbooks, and other employment documents should indicate that all work created by employees within the scope of their employment is to be treated as work-for-hire, owned by the employer. All contracts and employment documents should clearly indicate that all employees and contractors expressly agree to assign to the employer all rights those parties may have in all works that they create as part of their employment.

Clear job descriptions

All job descriptions and descriptions of duties set for contractors should be precise and accurate. Those descriptions should also be updated regularly to ensure that they accurately reflect the actual functions of the employees and contractors.

Manage licensed material

To manage copyrighted material obtained from other parties, your enterprise should take the following actions: (1) always use written licenses, (2) designate specific personnel who will be in charge of monitoring license compliance, (3) perform periodic audits of the inventory of licensed material and of license compliance, and (4) establish clear compliance procedures that licensors can follow when they have concerns about potential license compliance problems.

Appendix 6B: Trademark management strategies

Domain names and other forms of computer system content are commonly protected by trademark law. To enforce its rights, and to reduce the risk of trademark law liability to other parties, computer system owners should follow these general guidelines.

Thorough trademark development planning

As trademarks are planned and developed, legal input (e.g., advice, trademark searches) should be sought to ensure that the marks selected are readily defensible and do not infringe on preexisting marks developed by other parties.

Registration

Trademarks should be registered with the U.S. Patent and Trademark Office. Foreign registration is also prudent if international use is contemplated.

Usage guidelines

Trademark rights are substantially dependent on consistent use of the mark by its owner (i.e., all marks must be used in commerce in exactly the same form as that which is specifically claimed as the trademark: same color, same type style). Usage guidelines should be developed and enforced to make sure that all marks are used consistently.

Record of use

For each trademark claimed, your organization should create a complete record of use. At least one sample representative of each commercial use of each mark (e.g., promotional, advertising, marketing material) should be retained in a secure, readily accessible location.

Monitor marketplace

Your organization should regularly review activity in the marketplace to monitor for misuse of your trademarks by licensees or use of confusingly similar marks. This review should include on-line searches, review of advertising, and monitoring of domain name registrations. Instances of apparent misuse should be investigated. When appropriate, legal action should be instituted in order to protect the trademark rights.

Mandatory use of licenses

In all instances when the organization authorizes another party to use a trademark, that authorization should be documented in a written license. The license should clearly define the limits of authorized use and the term of the authorization. Periodic compliance audits should be conducted to monitor performance by licensees.

Appendix 6C: Managing trade secrets and other proprietary material

Ineffective management of trade secrets and other forms of confidential material can result in competitive harm, legal liability to other parties, and a loss of rights by the computer system owner. To avoid those difficulties, system operators should follow these basic principles for effective trade secrets management.

Identify confidential material

All trade secrets and other forms of confidential material should be identified and inventoried.

Access and use controls

Procedures to restrict access and control use of the proprietary material should be implemented. Those controls must be adequate to ensure that legal rights as to trade secrets are preserved and that all legal requirements for protection of the trade secrets of other parties are met. The controls should be updated and tested on a regular basis.

Confidentiality agreements

Contracts defining the terms of protection should be executed for all trade secrets that are shared with employees and other parties. When the organization receives trade secrets from other parties, confidentiality agreements should also be executed.

Enforcement

System use should be monitored to ensure compliance with all legal requirements and all confidentiality practices. Specific penalties for failure to comply with trade secrets protection practices should be clearly defined and consistently enforced. Failure to comply with trade secrets protection practices should be specifically identified as a basis for termination of employment.

Audits

Audits should be conducted on a regular basis to ensure effective compliance with the trade secrets management practices. These audits should test for effective implementation of the practices and should provide a review of the inventory of confidential material held by the organization.

7

Preserving E-Commerce
Transaction Integrity

As more commercial activities move on-line and into electronic form, parties to those electronic transactions require, just as they do for in-person transactions, effective protection of the integrity of the transactions. Failure to preserve e-commerce transaction integrity can lead to adverse economic consequences by undermining user confidence, thus impeding the expansion of e-commerce. Failure to preserve e-commerce transaction integrity can also generate legal liability for all participants in those transactions. To preserve the integrity of electronic transactions, certain formalities established by law must be satisfied. This chapter addresses some of those formalities and the basic legal issues associated with electronic commercial transactions.

Electronic commerce transaction integrity requires certain basic elements. It requires that all information associated with the transaction (e.g., payment information) be protected from unauthorized interception or alteration. Identity of the parties participating in the transaction should be verified, and the authenticity of all goods and information that are essential to the transaction should be established. The transaction should be processed in a manner that creates a binding, legally enforceable agreement, and thus prevents repudiation of the agreement by any of the participants. Finally, the transaction should be properly documented, creating a legally valid trail of records to verify that the transaction was fully executed, and that documentation should be securely stored to ensure that it has not been altered. In this chapter, we will discuss some of the legal rights and obligations associated with the protection of these fundamental elements of standard electronic commerce transactions.

Key elements of electronic contracts

Contracts, the legally enforceable exchange of promises at the heart of commercial transactions, provide an important set of legal rights and obligations associated with e-commerce. Contract law requires satisfaction of several basic criteria before a legally enforceable contract can be formed. If parties fail to meet these legal criteria, the law will not enforce the terms of their agreement. Under those circumstances, if a party to the agreement fails to perform its promises, the other party will have no legal recourse. Electronic contracts and transactions must meet these legal formalities if they are to be enforceable.

The parties to a contract must all have the legal capacity to enter into a contract. Legal capacity requires that the parties to the contract all be adults and that they all possess mental capacity sufficient to enable them to understand the terms of the contract and the fact that the contract makes them legally responsible to perform in a manner consistent with the contract terms. If one of the parties to a commercial transaction does not have legal capacity, in many instances the promises made by that party will not be enforced by law.

The parties must all understand the terms of the contract and must intend to be bound by the contract. Understanding of the terms and willingness to be bound by those terms is known as *assent*. Effective assent can only be established when the parties have dealt with each other in a fair manner, without fraud or misrepresentation of facts important to the transaction.

There must be a valid offer for the transaction and a corresponding acceptance of that offer. The offer is a specific invitation from the seller to the buyer, describing the product or service and defining the key terms of the transaction (e.g., price). The acceptance is an acknowledgment by the buyer indicating that the terms presented in the offer have been received, are understood, and are acceptable to the buyer. Offers and acceptance can be in specific written form, in oral form or expressed through the actions (conduct) of the parties, but they must both be present if there is to be a valid, enforceable contract.

The parties must exchange promises that have value. This requirement is known as the requirement for *consideration*. Consideration does not necessarily require that money change hands as part of the contract transaction. Instead, the consideration requirement makes it essential that each party to the contract must be obligated to perform an action that is either beneficial to the other party or detrimental to the party who performs the action.

The challenge facing electronic commerce operators and public policy makers is to find ways to apply the basic rules of contracts and commercial transactions to electronic transactions. The current legal trend in many jurisdictions is to grant electronic agreements and transactions the same legal status as that granted to in-person or paper contracts. The legal trend is to create rules that prohibit parties from challenging the validity of an e-commerce transaction or document simply because it is in electronic form. However, most of these rules continue to allow parties to challenge the validity of an electronic contract or document using the same legal arguments that they can raise to challenge the validity of a paper or oral contract. For example, while a party to an electronic contract will not be able to argue that the contract is invalid simply because it is in electronic form, that party could challenge the validity of the contract by demonstrating that there was no valid consideration supporting the contract.

The primary lesson for e-commerce participants is that they must structure their electronic transactions and supporting documentation to create a verifiable record that shows that the basic legal elements of contract law have been satisfied. Even if their transactions are processed entirely in electronic form, the records documenting those transactions must be structured so that they demonstrate that the parties, in fact, created a binding contract, based on valid contract capacity, mutual assent, proper offer and acceptance, and effective consideration. This record for all e-commerce transactions is essential to ensure that those transactions will be legally enforceable.

Digital signatures

E-commerce transaction integrity requires an effective means to verify the identity of the parties involved in the transaction and to demonstrate that the parties want to be bound by the commitment (i.e., that they assent to the commitment). In traditional commerce, these steps are accomplished through use of a signature. When we sign a contract, our signature is proof of our identity and verification of the fact that we assent to the transaction and agree to be legally bound by its terms. Electronic commerce must develop a comparable means of establishing identity and confirming assent to the contract commitment.

To accomplish the legal functions of establishing identity and demonstrating assent, various forms of digital signatures have developed. In the early days of electronic commerce, the legal validity of digital signatures was commonly challenged by parties who wanted to be relieved from their obligations under electronic contracts. Parties commonly argued that the use of digital signatures did not adequately prove the contract law requirements of proper identification and assent to be bound by the contract. Those arguments were, at times, successful in courts in various jurisdictions. This created a climate of uncertain enforcement for electronic contracts, and that uncertainty was a threat to e-commerce expansion.

To help facilitate electronic commerce, many jurisdictions have now enacted laws that specifically recognize the legal validity of digital signatures. In the United States, for example, many states have enacted versions of UETA and the federal government enacted E-Sign. The U.S. laws recognize the legal validity of electronic signatures and they make it clear that electronic contracts and other forms of electronic documents cannot be challenged based solely on the fact that they are in electronic form. The laws do, however, permit electronic signatures to be challenged in the same ways that the validity of traditional signatures and documents can be challenged (e.g., forgery, duress). Both UETA and E-Sign do not specify a single system or technology for valid electronic signatures. Instead, they leave that decision to the parties involved in the transaction. Any reasonable system for an electronic signature that is acceptable to the transaction parties will be valid.

Many other countries have also enacted laws that give legal enforceability to electronic signatures. The European Parliament adopted its Directive for a European Community Framework for Electronic Signatures. Bermuda made electronic signatures valid through enactment of its

Electronic Transactions Act, and Malaysia has implemented its Digital Signature Act. Israel implemented its Electronic Signature Law, and Tunisia has recognized digital signatures under its Electronic Exchanges and Electronic Commerce Law. Venezuela enacted the Law on Data Messages and Electronic Signatures, and the United Kingdom passed its Electronic Communications Act. Digital signatures have also been given legal force in Germany through the Digital Signature Law, in South Korea under its Digital Signatures Act, in India under the Information Technology Act, in Singapore under the Electronic Transactions Regulations, and in Argentina through the Digital Signature Legislation.

As electronic signatures gain full legal validity in more jurisdictions, the primary legal issue associated with their use will likely be the need to manage their security. Digital signatures can be used to create legally binding commitments; thus, everyone must be sure to implement security measures adequate to protect the integrity of the digital signature systems. Parties that create, operate, maintain or manage digital signature products (e.g., hardware or software) or digital signature services face legal liability from the users of those systems if the products or services they provide fail to operate properly or if there are security breaches that result in misuse of the signatures.

An interesting additional source of potential legal liability associated with digital signature use involves disclosure of personal information as part of the digital signature process. Some observers have described this potential privacy problem as *information leaking*. There is increasing concern that commercial parties will have incentive to encourage the providers of digital signature systems to structure those systems so that they transfer more personal information about the signature owner than the minimum amount necessary to make an effective verification of identity.

For example, a business may find it valuable to obtain the birth dates of its customers as part of the identity verification process, and thus might urge electronic signature system developers to include that information in the digital signature. From a strict perspective, that additional personal information is not necessary to make an identification of the individual or to verify the person's assent to the transaction and capacity to conduct the transaction. Instead of requiring an actual birth date, the electronic signature could simply have included verification that the individual was more than 21 years of age. The personal information has thus leaked into the transaction but is not necessary for the transaction.

As concern about transfer and use of personal information increases, it is likely that this type of information leaking as part of the electronic signature process will draw more legal attention. Prudent developers and operators of electronic signature systems and businesses that make use of those systems should be cautious about the amount of personal information that they integrate into the systems. All personal information that becomes part of the electronic signature process must be managed in compliance with legal protections provided for that type of information. A developer or a user of an electronic signature system can be liable for information privacy or security violations caused by that system. Those system developers and users should strive to minimize the amount of personal information transferred by the systems and to structure the systems so that they are in full compliance with all legal obligations as to the personal information that is used by the systems. Failure to comply with information privacy requirements, as to personal information distributed in conjunction with electronic signatures, can create legal liability for the parties that collect, distribute, or use that information.

Certification authorities

To provide additional security to electronic transactions, some e-commerce participants now seek an independent party to verify the identity of the participants in those transactions. Many jurisdictions have long made use of a neutral party known as a *notary public*. Commonly certified by governments, these notaries performed the function of confirming the identity of parties involved in commercial and civic transactions. An individual would sign a contract or other official document in the presence of the notary, and the notary would also sign the document, legally attesting to the fact that the signer was, in fact, the person he or she claimed to be, and that the person had, in fact, actually signed the document. At a simple level, certification authorities perform the notary public role for electronic transactions.

Certification authorities have been legally recognized in several jurisdictions. For example, India's Information Technology Act gives legal recognition to the certification authority function. Germany's Digital Signature Law performs a similar function, as does Singapore's Electronic Transaction Regulations, and South Korea's Digital Certification

Authority Regulations. Argentina's Digital Signature Legislation also gives legal status to certification authorities. In addition to providing legal recognition of the certification authority function, these laws also provide for government approval and registration of certification authorities. Essentially, these rules establish the standards required for government approval of private certification authorities and they define the procedures to be applied by government as it determines which applicants will be selected as qualified certification authorities.

One important legal issue that has not yet been fully resolved in many jurisdictions is the extent to which a certification authority will be legally liable if it makes an error in a certification. For instance, if an authority certifies the identity of a buyer in an electronic transaction, but the buyer can demonstrate that he or she did not, in fact, conduct the transaction, it is unclear the extent to which the certification authority that made the error could be held accountable for any damages caused by the misidentification. Some jurisdictions, Singapore for example, protect the certification authority from liability associated with errors provided that the authority followed the operational regulations established by the government as requirements for all certification authorities. Other jurisdictions are, however, unclear as to the extent of the authority's potential liability for misidentifications. It is possible that the authority could be liable for damages suffered by both the seller, in the example above, and by the party who had its identity stolen.

Another source of potential legal liability for certification authorities is a breach of the authority's security. If, for example, the authority performs its certification function through use of public key encryption systems, and if that system is somehow compromised, it is possible that the authority may be liable for all damages suffered by its clients as a result of the security breach. This risk is a particularly important one for certification authorities as their systems and databases are likely to be highly attractive targets for criminals, and there will surely be instances when a certification authority's system will be compromised by a deliberate, malicious attack.

Certification authorities face significant potential legal liability. Their clients will be all parties to electronic transactions who rely on their certifications, and if there are any failures in the certification process, those clients will likely take legal action against the authority to recover compensation for damages suffered as a result of the failure. Parties considering performing the certification authority role should consider the significant potential legal liabilities associated with that role before making a final decision to accept that responsibility.

Perhaps the most appropriate policy approach to the certification authority process is one that makes use of government regulation of those authorities. The jurisdictions that have established government-enforced selection and operating criteria for certification authorities appear to be following the best track. Under this system, qualified certification authorities would receive licenses issued by the government. Licensed certification authorities operating in a manner consistent with the government requirements would be immune from legal liability (or would have their liability capped at a maximum level by the government) for errors or security breaches. Unlicensed certification authorities and licensed authorities operating in a manner inconsistent with the government requirements would face unlimited legal liability for their failures. To supplement this process, governments could require all certification authorities (both licensed and unlicensed) to obtain and maintain specified levels of insurance coverage or to be bonded, to help ensure that parties who are harmed by certification failures will be able to recover some amount of compensation.

Payment processing

Processing of payments associated with electronic commerce transactions commonly involves several parties. In addition to the buyer and the seller in the transaction, there is generally a payment intermediary (e.g., a credit card company or some other financial service provider). The relationships among those parties are substantially affected by legal rights and obligations.

Contract relationships exist between the payment intermediary and both the seller and the buyer. Failure by any of those parties to perform in a manner consistent with their contract obligations can create a liability for breach of contract. For example, a buyer who fails to pay a legitimate debt owed to an e-commerce seller can be liable for breach of contract to the seller and to a payment intermediary.

Credit card companies and other financial service providers performing the payment intermediary role in the United States are also regulated by specific rules applicable to the provision of financial services to consumers.

The relationships between those enterprises and e-commerce buyers who qualify as consumers are substantially governed by specific financial services rules and consumer credit protection regulations. Financial service providers functioning as e-commerce payment intermediaries must comply with those rules as they process consumer payments in support of e-commerce transactions. Thus, an e-commerce buyer who disputes a charge that has been assessed against his account has a set of consumer rights that he can exercise (e.g., the right to withhold payment and force the credit card company to investigate the dispute).

One of the important public policy aspects of the e-commerce payment processing setting is the issue of who bears the risk of loss for repudiation of e-commerce transactions (e.g., in the event of an unauthorized on-line purchase). In jurisdictions such as the United States, that risk of loss is currently borne by the e-commerce seller, as a result of the substantial use of credit cards for e-commerce payments. Under federal rules in the United States, credit card users have limited liability for unauthorized use of their accounts, provided that those users follow the mandatory procedures for promptly disputing contested charges to their account and for cooperating with the credit card company as it investigates the dispute. The credit card companies are also in a position to limit their loss in the event of misuse of the card, based on their contractual arrangements with the sellers. In this environment, the seller bears much of the risk of loss associated with unauthorized credit card transactions.

Increased use of e-commerce payment systems other than credit cards can alter the sharing of risk of loss associated with e-commerce. For example debit cards and forms of electronic or digital cash provide alternatives to credit card payment systems for on-line transactions. Both of those payment systems provide for more immediate payment to the seller than that provided by the credit card system. Payment is processed more quickly as it generally involves a transfer of funds from a prepaid account. With these systems, risk of loss for an unauthorized purchase is borne, to a greater extent, by the buyer than is the case with credit card transactions. If these payment systems become more widely adopted to support e-commerce transactions, there will likely be a corresponding rise in legal conflicts between buyers and the payment processors as those buyers pursue legal claims against the payment processors to recover unauthorized payments.

Transaction documentation

Parties who participate in electronic commerce transactions should create and maintain effective documentation to record the history of each transaction. Some forms of transaction documentation are specifically required by law. For example, if the electronic transaction involves a sale of a product for which sales tax must be assessed, the seller must retain documentation identifying the buyer and the jurisdiction of the applicable tax. That electronic sale also triggers the obligation on the part of the seller to collect the appropriate amount of the tax from the buyer and to remit that amount to the proper tax authority.

Other forms of transaction documentation are important in order to preserve legal rights of the parties involved in the transaction. For instance, if the transaction involves the creation of an electronic contract, proper documentation of the history of the interaction between the parties is essential in order to demonstrate that the agreement between the parties forms a valid contract and is thus legally enforceable. Failure to create and retain an electronic record documenting that the required elements of a binding contract (capacity, assent, offer, acceptance, consideration) exist, could result in an unenforceable agreement. If the agreement fails to meet the required legal formalities, either party could avoid performance of their promises, and the other party would be unable to exercise legal rights to compel performance or to recover monetary compensation for damages caused by the failure to perform.

A critical aspect of proper transaction documentation is protection of the security of the records. To have legal value as evidence, there must be a verifiable means to demonstrate that the documents accurately reflected the transaction at the time of the transaction, and were not altered in any way after the transaction. In a previous chapter, we discussed the importance of effective protection of electronic records. Electronic commerce transaction documentation is an important form of legal records that should be managed using the highest level of care, as described in that previous chapter.

Transaction security

Parties involved in electronic commerce transaction processing may face legal liability if they fail to provide adequate security for those transactions.

For example, if an unauthorized party intercepts payment information, such as credit card account numbers, or protected personal information, such as addresses or phone numbers during or after a transaction, the party who processed that transaction will likely face legal claims raised by the party whose information was compromised.

Private parties harmed by a breach of security can raise tort law claims to recover damages in the United States. To recover damages for a tort claim, the party must demonstrate that the transaction processor had a legal duty to protect the security of the transaction content and that the processor failed to meet that duty, resulting in harm to the party raising the claim.

Failure to provide adequate electronic transaction security can also lead to liability for breach of contract. If contracts (including statements of terms of service) between the transaction processor and the parties to the transaction establish any obligations as to transaction security, failure to meet those obligations could constitute a breach of contract. If the party can demonstrate that the contract obligation that was not fulfilled was significant, that party can recover compensation for the damages caused by the contract breach.

If the material that was the subject of the security breach qualifies for special protection under specific laws or regulations (e.g., certain forms of personal information), an e-commerce enterprise that failed to provide the security may face legal liability under those specific legal requirements. For example, federal law in the United States requires use of reasonable security measures to protect the privacy of personal financial information of consumers and personal medical/health information. If such information was intercepted in the course of an on-line commercial transaction, the party responsible for processing that transaction would likely face legal claims based on the privacy regulations.

UCITA

A uniform statute to govern the sale of computer software and information products is being considered for adoption in many of the states in the United States. UCITA has been adopted in two states (Virginia and Maryland) and is under consideration in more than 10 other states. The terms of UCITA will affect many e-commerce transactions. UCITA applies to the sale of computer software and information products. For purposes of the act, information products essentially include any type of content that can be stored or distributed on computers. This definition brings material such as text, sound recordings, images, and motion video within the scope of UCITA, when that material is stored or distributed using computers. UCITA has broad scope, affecting transfer of a wide range of digital content.

UCITA continues to recognize computer software and information products as intellectual property, subject to intellectual property licenses. Both shrinkwrap licenses and on-line *click-through* licenses are given full legal value under UCITA, provided that those licenses contain terms that are reasonable and consistent with standard market practices. The drafters of UCITA indicate that the act does not in any way reduce the rights and obligations established by traditional intellectual property law.

Based on specific substantive concerns, UCITA has been opposed by several different groups. For example, parties who make significant use of copyrighted material under *fair use* provisions (e.g., libraries, schools) have commonly opposed adoption of UCITA fearing that the statute would reduce their access to fair use materials. In part, this concern is based on a belief that copyright owners might circumvent the fair use exemption granted by copyright law by making access to their property subject to UCITA commercial licenses instead of traditional copyright licenses. The fair use advocates are concerned that licenses established under UCITA are not required to accommodate the fair use exemption; however, UCITA proponents contend that the fair use exemption would continue to exist, even under UCITA licenses.

Another objection raised against UCITA is that it favors the sellers of software and information products. One example of this purported bias is the provision in UCITA that enables sellers of software and information products to engage in "self-help" measures to ensure that buyers of the information products comply with the terms of the license. For instance, a software developer could include a disabling function that the seller could activate if the buyer fails to meet the license terms. UCITA would permit use of these self-help measures by sellers, and certain groups representing likely users of the products covered by UCITA oppose that principle.

UCITA illustrates a growing trend toward the development of commercial laws to help make the terms associated with the transfer of digital content more uniform. Many observers believe that a more uniform set of commercial rules applicable to the sale and lease of information products will help to facilitate the rapid expansion of electronic commerce. We will likely continue to see more jurisdictions adopt UCITA and other standardized commercial transaction rules to help make e-commerce transactions easier to execute.

Parties involved in the transfer of software and other forms of information products, as either sellers or buyers, should be mindful of UCITA and other applicable commercial transaction laws. They must recognize that transactions that were formerly governed entirely by intellectual property law may now also be affected by various laws applied to commercial transactions and the sale of goods. The legal requirements associated with those transfers now include traditional intellectual property law, the specific contract terms associated with the intellectual property licenses negotiated by the parties, and the commercial transaction rules established by UCITA and other information transaction laws.

We will likely see increasing reliance on all of these different sets of legal requirements in conjunction with the transfer of information products. For instance, copyright licenses will continue to be the primary vehicle for the transfer of computer software, but those licenses will now commonly include provisions that make use of UCITA terms and concepts. It will be common to see software licensors include in the standard terms of their licenses choice of law provisions that apply the law of states that have adopted UCITA, thus making UCITA's transaction rules applicable to supplement the specific terms of the license. Standard licenses will now likely specifically reference UCITA and expressly indicate that UCITA terms are incorporated into the terms of the license by reference. In this evolving commercial setting, traditional intellectual property rights will be supplemented by commercial law principles with regard to the full range of information products.

Also recognize that the terms of UCITA enacted in different states are likely to vary slightly. One example of variation is the issue of jurisdiction. Some state law versions of UCITA attempt to give the local courts in the state jurisdiction over commercial disputes brought under the law. For example, the version of UCITA enacted by Maryland appears to give Maryland courts jurisdiction over all UCITA disputes when either the buyer or the seller of the information product is a Maryland resident. Some sellers are likely to try to override this provision of the law by stating in their license agreements that disputes under the license must be handled in a specific jurisdiction; however, it is likely that those license provisions will fail when they conflict with state law UCITA statute terms.

In the case of the Maryland version of UCITA, for instance, a seller of an information product located outside of Maryland would likely be forced to litigate in Maryland courts in the event of a dispute with a buyer of the product who was a Maryland resident, even if the specific license involved stated that all disputes under the license were to be handled in courts in a different jurisdiction. Buyers and sellers of information products that come within the scope of UCITA must, accordingly, be mindful of substantive differences in the content of the versions of UCITA enacted in different states, as those differences can have a significant impact on the scope and application of the terms of the statute.

Notices and click-through agreements

Two essential elements of proper documentation of electronic commercial transactions are effective use of on-screen notices and appropriate click-through agreements. Notices and click-through agreements provide the basic documentation necessary to establish legally enforceable commercial transactions. Parties who conduct on-line commercial transactions should be particularly careful to make sure that their transaction systems provide notices and agreements that are comprehensive and clear, as those documents will be the core of their electronic commercial records.

On-screen notices are important to establish that certain basic elements necessary for a binding commercial transaction have been effectively met. For example, the notices commonly indicate that the party conducting the transaction has contract capacity (e.g., is an adult) or that the transaction is legal in the jurisdiction in which the party conducting the transaction resides. These notices should always be used in conjunction with a click-through agreement to give them the greatest chance of being deemed to be enforceable in all relevant jurisdictions.

Click-through agreements are basically evolving into the electronic equivalent of the traditional contract. The terms and conditions associated with these contracts are presented on-screen. The party receiving the offer is given an opportunity to review those terms, and if that party finds the terms to be acceptable, it indicates acceptance of the offer by clicking an on-screen icon designating "I accept" or "I agree." Properly structured, click-through agreements provide the mechanism to meet the basic formalities for a binding agreement required by contract law. Click-through agreements can provide proof of a valid offer and acceptance. When used properly with on-screen notices, they can also provide evidence of contract intent (i.e., assent) and the contract capacity of the accepting party. Click-through agreements and associated on-screen notices should be carefully structured to ensure that they establish binding, enforceable contracts.

Taxation

Electronic commercial transactions can generate legal obligations associated with the collection of sales tax or the payment of use tax. The subject of taxes applied to electronic commerce transactions remains unresolved and volatile, at present, yet there are already certain tax collection and payment obligations in effect. The current trend is for governments to treat on-line sales in the same way that they treat other forms of distance selling (e.g., sales via mail, telephone, facsimile) for sales and use tax purposes. Merchants conducting electronic commercial transactions with retail consumers must make sure that their e-commerce systems are structured in ways that enable them to comply, as necessary, with sales tax collection obligations. Failure to do so can result in a tax liability for those merchants. Merchants who have an obligation to remit sales tax payments to a state for specific transactions will be liable for those amounts. Thus if the seller does not collect the appropriate tax from the buyer, the seller will, in many cases, be accountable to the state for the amount of sales tax owed.

Many states in the United States impose a legal obligation to collect and remit sales taxes for certain products sold to consumers who are residents of those states. Under current federal law in the United States, each state is permitted to require a seller of qualifying products who is located outside of the state to collect and remit to the state sales tax assessed against purchases made by residents of the state. This collection obligation can currently be imposed, however, only on sellers who have some type of physical presence in the state assessing the tax (e.g., an office, sales staff, a manufacturing facility). Sellers located outside of the state who do not have a presence (nexus) in the taxing state are not presently required to collect sales tax; however, the consumers who make such purchases are required to pay use tax for those products.

Use tax is the mirror image of the sales tax. Most states set their use tax rate equal to their sales tax rate. The use tax is paid directly by the purchaser of the product to the state taxing authority, often as an adjunct to the state income tax filing. The seller is not obligated to collect or remit the use tax. Sales and use taxes imposed by the various states in the United States are currently applicable to all distant sales, including sales made using the Internet. At present, however, the U.S. federal government has imposed a moratorium on new state taxes applied exclusively to sales made using the Internet. Note that this moratorium does not prevent states from assessing already existing sales and use taxes that apply to all distance sales, including those on the Internet. The moratorium was intended to provide the U.S. government with an opportunity to study the potential impact of taxes on Internet sales and to develop plans for an appropriate tax system. The debate in the United States continues, at the time of this writing, and no clear Internet sales tax strategy has yet been adopted.

Diverse models for electronic commerce sales tax are currently under consideration. Some advocate that the local sales tax of the buyer should be applied to all distance transactions, regardless of whether the seller has a physical presence in the jurisdiction in which the buyer resides. Others argue that there should be a specific tax applied to all distance transactions, thus avoiding a situation in which sellers must comply with thousands of different local tax rates. There are others who contend that all distance purchases should be exempt from sales tax. It appears most likely that sales and use tax will continue to be applied to all distance purchases, including those conducted using the Internet, for the foreseeable future.

Electronic commerce also raises issues associated with income tax. Merchants who sell products to buyers located in other jurisdictions commonly generate income in those jurisdictions. In many instances, e-commerce operations can thus create income tax liability in the various jurisdictions in which the merchant is selling its products. To the extent that your business is generating earnings in other jurisdictions through its electronic commerce activities, it is important that you become aware of the income tax requirements of those jurisdictions. Failure to report income and pay income taxes, as required, can lead to significant legal liability.

Electronic commerce operations also generate potential tax liability associated with telecommunications and information services. Some jurisdictions, for example, apply taxes to telecommunications services or equipment. Others apply specific taxes to the purchase of information products or services (e.g., database content). Certain e-commerce functions can fall within the scope of these specific taxes; thus merchants and service providers in the digital marketplace should be sure to seek appropriate tax-law guidance in support of their operations, in all jurisdictions in which they conduct business.

Contraband and illegal products

A key aspect of the integrity of electronic commerce transactions is the need to ensure the legality of the goods and services that are the subject of the sales. An on-line merchant who sells regulated content must ensure that its e-commerce systems prevent illegal sales. Compliance with legal requirements associated with sale of regulated products and services is a critical component of overall commercial transaction integrity.

There are many examples of on-line sale of illegal content. In some instances, the sale involves a product that is completely banned. An example of this situation was the sale of hate-based content, prohibited under French law, to French consumers using on-line auction systems, including that of Yahoo!. In that case, French authorities prosecuted Yahoo! and received a French court judgment ordering Yahoo! to block those sales. The French court asserted jurisdiction over the U.S. company Yahoo!, based on the on-line sales to consumers located in France.

In other instances, the sale of certain products to specific groups of consumers may be illegal, but the sale of those same products to other consumers in the jurisdiction may be fully legal. For example, in the United States it is legal to sell tobacco products, alcohol, and sexually oriented material to adults, but illegal to sell those same products to minors. For these products, a seller must implement a commercial transactions system that enables the seller to distinguish among the different groups of consumers, permitting sales to those authorized to buy and blocking sales to those who are not permitted to purchase the products.

Parties selling goods or services that are either prohibited or restricted must make sure that their electronic transaction processing systems are adequate to ensure compliance. If their e-commerce systems are not able to monitor and control the sales, the operators of those systems will face legal liability. That liability can take the form of fines or prison time based on the sales of the illegal products. A merchant who fails to comply with legal restrictions on sale of certain goods can also be prohibited from selling those products in the future (e.g., sellers of alcohol products can lose their license to sell those products if they fail to comply with restrictions on sales to underage buyers).

Transaction integrity for the sale of regulated goods and services may also require controls on the content that is being distributed. For example, an on-line seller of prescription drugs must comply with all regulations applied to the sale of those drugs. Most jurisdictions require that prescription medications can only be distributed by licensed pharmacists. An on-line seller of prescription drugs must, therefore, ensure that its products are being distributed only by appropriately licensed pharmacists. An on-line system to process those sales must, accordingly, be structured in a manner such that the requisite safeguards applicable to the management of the product and the transactions are met. In these cases, the e-commerce transaction system must be designed and operated in ways that provide assurance that the product controls required by law are satisfied in each transaction.

Dispute resolution

Commercial transaction disputes can be resolved by formal courts (through traditional litigation) or by private parties or organizations (e.g., mediators, arbitrators). Increasingly, e-commerce participants are relying on private systems for dispute resolution instead of the traditional court system. Parties opting for that approach recognize that these alternative dispute resolution systems are often less expensive to use and are frequently able to resolve disputes more quickly than courts. Proponents of traditional litigation in courts note, however, that decisions reached by courts can often be enforced more effectively than can the judgments made by private arbitrators or mediators. Enterprises involved in electronic commercial transactions should seriously consider specifying in advance which type of system will be used to resolve future disputes.

The most commonly used alternatives to court litigation are mediation and arbitration. Both of these systems generally involve the use of one or more neutral parties to hear the facts of the dispute and to work with the parties to resolve the dispute. A mediator commonly works with the parties to help them try to develop their own settlement that is acceptable to every-one involved in the dispute. An arbitrator tends to function more like a tra-ditional judge, listening to the arguments of the parties, evaluating the evidence presented, and ultimately issuing a decision or order to resolve the matter.

One of the best current examples of thriving on-line use of alternative dispute resolution systems is the domain name dispute system. Arbitration systems to resolve domain name controversies are currently operated by several parties, including WIPO. These systems have become highly popu-lar as they can be used relatively quickly and at low cost. They also have the advantage of being more convenient when, as is often the case with domain name disputes, the parties involved are located in different countries.

Many different private groups provide third-party dispute resolution services for commercial disputes. For example, the International Chamber of Commerce provides those services at the international level. In the United States, the American Arbitration Association offers third-party neutrals. Organizations that have ongoing business relationships with each other often mutually agree on appropriate private parties (e.g., lawyers, consultants, former judges) on their own, to serve as neutrals in the event there is a commercial dispute.

Arbitration and mediation groups are beginning to make use of on-line dispute resolution systems, although those systems remain, at present, in the very early stages of acceptance. For example, the WIPO system for domain name dispute resolution (and several of the private domain name dispute-resolution systems) can be conducted almost entirely in electronic format. The claim is raised, the panel of arbitrators is selected, the cases are presented, and the opinion of the arbitrators is rendered in electronic form. This electronic process helps to reduce the time and expense required to complete the process. It should be noted, however, that the judgments delivered by these electronic systems have not yet been fully tested in the courts; thus we are not yet fully certain as to how effectively their decisions will be enforced.

A party involved in electronic commerce should seriously consider use of private dispute-resolution systems as its standard approach to address e-commerce transaction conflicts. If the enterprise chooses to use private dispute resolution, it should clearly express that decision in its on-screen notices and all associated contracts. The party should be specific about the type of dispute resolution to be used (i.e., mediation or arbitration) and it should clearly describe how the neutral party will be selected. Finally, the party should indicate how the decision of the neutral party will be enforced. If properly structured, use of alternative dispute-resolution systems can be particularly helpful for enterprises engaged in electronic commerce as they can provide an efficient and less expensive method to resolve commercial disputes involving parties located in many different jurisdictions.

When considering use of these alternative dispute-resolution systems, it is important to focus on at least two criteria as the system is planned. The first essential criterion is the enforceability of the final judgment. If you choose to use arbitration or mediation, it is important that the system be structured and operated in a manner that ensures that the final results of the process can be effectively enforced. It will be a waste of resources if the system leads to a conclusion that cannot be effectively enforced. Also, it is important to make sure that the process is structured so that the neutral party who leads the process has the experience and expertise necessary to render a fair and reasonable decision. If you are going to use arbitration or mediation, make sure that the process for selection of the arbitrator or mediator is adequate to generate an appropriate party or parties. The neutral parties selected to lead the dispute resolution process are the key to the success of that process.

Appendix 7A: Creating enforceable commercial contracts

The law requires that certain formalities be satisfied before a commercial contract will be enforceable. Those legal formalities apply to contracts in all forms, including electronic or on-line agreements. Parties engaged in electronic commercial transactions must make sure that their e-commerce agreements comply with the following legal standards.

Parties have legal capacity

To create an enforceable contract, all parties to the agreement must be adults and have adequate mental capacity to understand that they are entering into a legally binding commitment. This criterion is easier to satisfy when in-person negotiations are involved instead of electronic communications and contracting. When the transaction is an electronic one, the best approach is to have the terms of the electronic agreement include *representations* (i.e., specific statements attesting to facts) from the parties indicating that they have the legal capacity to enter into a binding contract.

Demonstration of assent

Use of some form of electronic signature or an "I accept" or "I agree" icon in a click-through agreement generally provides a satisfactory demonstration of the intent of the parties to be bound by the agreement.

Valid offer and acceptance

Use of an on-screen notice and electronic contract that specify the important terms of the transaction provide a valid contract offer. Acceptance can be validly provided through use of electronic signatures or an "I accept" icon. For a click-through agreement, extra confirmation of acceptance can be provided by using an additional screen (one that appears after the "I accept" icon has been selected) with text that confirms to the user that a contract has been executed and permits the user to print a hard copy of the confirmation screen as a receipt.

Exchange of consideration

To create a binding contract, the parties must exchange promises that require them to incur some detriment (e.g., pay money). The consideration for the agreement must be clearly defined in the agreement. It should be described in a way such that a third party who was not involved in the transaction would be able to identify the consideration based solely on a review of the contract.

On-screen notices

On-screen notices can be used to help explain terms and provide information to assist a party to make an informed judgment about an agreement. Note, however, that if a formal contract (electronic or otherwise) is used, all relevant terms associated with the transaction should be clearly expressed in the contract itself. Disclosing that information in a separate notice will not make it a part of the contract, unless the contract specifically states that the terms contained in the notice are incorporated by reference into the contract.

Click-through agreements

Click-through agreements can create binding contracts, provided that they satisfy all of the basic contract law requirements (e.g., offer, acceptance, consideration). Click-through agreements provide efficient and effective electronic contracts. Note that if click-through agreements are used, accurate copies of each executed agreement must be retained as part of the secure documentation for the on-line transactions. You must be able to produce accurate copies of each fully executed agreement and all of the material terms of each of the agreements must be clearly presented in the copy.

Shrinkwrap agreements

Shrinkwrap licenses are also legally binding, provided that the terms they contain are reasonable. In the past, some jurisdictions were reluctant to enforce shrinkwrap licenses because the party entering into the contract was generally unable to read the terms of the agreement before accepting them. Most jurisdictions now accept shrinkwrap licenses if the terms of the licenses are consistent with the standard terms included in licenses for similar products. The theory is that a license containing commercially standard terms will not surprise a consumer and will not place an unfair burden on that consumer.

Choice of jurisdiction

Contracts for e-commerce transactions should specifically identify a choice of law to be applied to interpreting the contract and resolving disputes associated with the contract. This involves identification of the jurisdiction whose laws will be applied and the legal forum to be used in the event of a dispute. If you want to use some form of dispute resolution other than litigation in a traditional court (e.g., mediation or arbitration), that choice should be specifically described in the contract.

Appendix 7B: Documenting e-commerce transactions

Parties who conduct electronic commerce transactions should be prepared to retain documentation describing and defining those transactions. The documentation is important for both commercial and legal reasons. Basic e-commerce transaction documentation should include the following information.

Executed contract

If a formal written contract has been executed (e.g., electronic contract, click-through agreement), a complete copy of that contract should be retained as the most important component of the transaction documentation. Whenever possible, it is best to create a formal, written contract, to provide a complete record documenting all of the essential elements of each transaction.

Verification of identity of parties

The means used to verify the identity of the parties to the transaction should be documented and made part of the transaction record. For example, if a digital signature is used, that signature and associated material should be retained as part of the transaction documentation.

Description of goods and services

The specific goods or services that are the basis for the transaction should be clearly described in sufficient detail to identify them and to permit effective delivery. This description should be included in the transaction record and used to confirm that the proper product is ultimately delivered. This description is also important in order to ensure that sales tax is collected, when required.

Transaction terms

The specific terms associated with each transaction (e.g., price, warranties) must be part of the transaction record. This information is essential in order to ensure that the contractual obligations associated with each transaction are effectively satisfied.

Transaction value

The sale price must be part of the transaction record. This information is particularly important in order to facilitate sales-tax compliance.

Proof of payment

Effective payment-processing systems should be part of each transaction, and each of those systems must provide accurate proof of payment status for the transaction documentation.

Product delivery

The transaction documentation should also include accurate records regarding delivery of the goods or services covered by the transaction.

Appendix 7C: Legal guidelines for use of electronic signatures

Organizations that develop or operate electronic signature technology and those that make use of electronic signature systems should be mindful of certain key topics of legal concern. Those important legal topics include the following.

Digital signature laws and regulations

The laws relating to digital signatures and validity of electronic records/ transactions in each of the jurisdictions in which your enterprise does business should be carefully reviewed and monitored. You must make sure that the digital signature systems you apply are consistent with the legal requirements in those jurisdictions, or you run the risk that the contracts and transactions supported by those systems may be unenforceable.

Basic contract law requirements

Electronic signature systems that you develop or use must be compatible with basic contract law requirements. Those systems must support accurate identification of the parties involved and effective communication of assent to, and acceptance of, the contract terms. If your system fails to meet these basic contract law requirements, it will not establish legally enforceable contracts.

System security

Prior to using electronic signatures, you should explore the extent that you may be held responsible for the security of the signature system. The key question is: To what extent is your organization legally liable for failures in the digital signature process that result in harm to other parties? The answer will likely vary in each of the jurisdictions in which you conduct business and will probably affect your decisions regarding which transactions you will support using electronic signatures, which type of signature systems you will use, and what security measures you will apply to protect the signature systems.

Information privacy

You should monitor the extent to which your electronic signature system involves use of personal information of individuals. It is likely that every digital signature system creates personal information records that carry legal protection requirements in at least some jurisdictions (e.g., the EC). To the extent that this is true for your system, you must make sure that the electronic signature records created during commercial transaction processing are effectively managed under your organization's information privacy practices.

Appendix 7D: Legal checklist for certification authorities

Certification authorities are playing an increasingly important role in electronic commerce, but the precise scope of that role remains uncertain in many jurisdictions. In that setting, certification authorities should consider the following key legal topics.

Applicable government regulations

All parties interested in performing the certification authority function should review and monitor relevant government regulations. The trend in many jurisdictions is toward government licensing of certification authorities. Compliance with government regulations will be necessary in order to obtain a license and compliance may also enable the authority to reduce its risk of liability in the event of a failure of performance.

Fraud and risk of loss

A key issue for certification authorities is the level of their responsibility in the event of fraud by another party in the transactions that they certify. Legal responsibility on this point remains unclear and will likely vary among jurisdictions. Certification authorities must be aware of the extent to which they will be held accountable for fraud on the part of other parties. Some potential authorities are likely to take the position that government regulation of the certification function will be commercially beneficial if that regulation results in a limit to the liability that can be incurred by certification authorities.

Liability for security breaches

Another key issue for certification authorities is the level of security they will be required to provide for the certification process. This issue is similar to the issue of certification authority liability for fraud. Some jurisdictions are establishing specific security practices to be followed by the certification authorities, but other jurisdictions do not set specific standards. Certification authorities must make sure that they comply with established standards and they must determine what liability they will bear if security is compromised. To the extent that government-established security standards are coupled with liability limits for all certification authorities who comply with those standards, many certification authorities will likely support that form of government regulation.

Information privacy obligations

The signature-certification process involves collection and distribution of personal information pertaining to specific individuals. That information is increasingly the subject of legal requirements. Certification authorities should generally try to minimize the amount of personal information that they collect or use, and to the extent that they handle such information, they must make sure that their information management practices comply with information privacy regulations in all jurisdictions in which they conduct business.

Insurance or bonding

Some jurisdictions may require certification authorities to be bonded or to carry insurance for their activities. The authorities must make sure that they comply with all such mandatory requirements, and even if there are no specific coverage requirements where they do business, the authorities may want to make a business decision to obtain such coverage, given the significant potential liability associated with the certification function.

Dispute resolution

Certification authorities may want to consider using arbitration or other forms of alternative dispute-resolution systems instead of traditional litigation in courts. The fact that they face potential liability in many different jurisdictions and that the disputes they are likely to encounter may be complex suggests that use of alternative dispute-resolution systems may be more advantageous for the certification authorities than traditional litigation. If they choose this approach, the authorities must make it clear in all of their contracts and terms of service that all disputes directed against the authority must be processed using arbitration.

8

On-Line Exchanges, Auctions, and Outsourcing

Three popular trends in the evolution of the digital marketplace raise particularly challenging security issues that merit special discussion. Increased use of on-line commercial (business-to-business) trading exchanges, Internet auctions, and outsourced electronic commerce functions carry the potential for significant legal liability. Parties participating in those e-commerce activities should carefully examine the rights and liabilities associated with those aspects of the electronic marketplace. Operators and users of commercial exchanges and on-line auctions have certain obligations with regard to the security of transactions, information management, and other operations. Providers of electronic commerce functions acting as contractors also have obligations with regard to the security of the functions they provide. Security for the operations of these e-commerce functions is essential to minimize the risk of improper use of the e-commerce systems. Failure to meet security requirements can result in legal liability for the parties involved in those failures. This chapter provides an overview of some of the most important liability concerns associated with security failures in these rapidly developing digital marketplaces.

Managing trade secrets in commercial exchanges

On-line commercial trading hubs or portals are, by definition, systems in which different, and sometimes competing, businesses share information. These trading exchanges can be structured in different ways. For example, some exchanges are developed by a single supplier (e.g., Cisco Systems or Dell) as a system to sell to many different customers. Other exchanges provide on-line marketplaces in which many different suppliers can sell to many different customers (e.g., e-Steel). Buyers can also create their own exchanges, as some governments and the automobile industry have done, for instance, to provide marketplaces through which they can attract suppliers and conduct their purchasing and procurement functions. Exchanges that focus on serving a single industry are commonly referred to as *vertical* exchanges, while those that match buyers and sellers in diverse industries are often described as *horizontal* exchanges. Examples of these electronic, business-to-business marketplaces include: Covisint (automobile industry), Globalnetexchange (retail industry), Transplace.com (freight transport industry), Forest Express (paper products), Petrocosm (oil and gas), New Health Exchange (health care), Rubbernetwork.com (rubber products), Metalspecrum (specialty metals), Transora (consumer products), Pantellos (energy), E2open (electronics products), and Omnexus (plastics).

On-line commercial exchanges thus require a challenging mix of information sharing and secrecy protection. Rigorous systems to manage distribution of information are particularly important for commercial trading hubs. Much of the benefit and efficiency derived from the on-line marketplaces comes from the ability to interconnect information and transaction processing infrastructures among companies that conduct business with each other. In that environment, substantial information sharing is necessary, yet that information will be shared in many instances among a diverse mix of suppliers, customers, and competitors.

Much of the information associated with the transactions processed by the exchanges will, however, be confidential. Information regarding quantities of products to be purchased and future pricing, for example, will often be commercially sensitive and treated as proprietary by participants in the exchange. Operators and users of the commercial exchange must make sure that the exchange infrastructure and its operational practices permit identification of confidential material and adequate protection of the security of that material. Failure to do so can result in legal liability, as the party whose confidential material was disclosed could sue the disclosing party for theft or misappropriation of trade secrets. Parties involved in the unauthorized disclosure of the trade secrets could also face liability, regardless of whether the disclosure was intended or unintended, for violation of civil and criminal statutes that protect the privacy of electronic communications. This on-line commercial setting thus creates an environment in which the operators and users of the exchange may face liability for unauthorized disclosure or other misuse of confidential material.

Assume, for example, that an on-line business-to-business trading exchange processes orders from various companies in an industry for component parts used to manufacture their products. During the process of negotiating and consummating the transactions for the component parts, information regarding pricing, number of units to be purchased, quality or performance specifications, and delivery requirements will be shared through the trading exchange. Some of that information will likely be considered confidential by the buyers and the sellers. If the sensitive confidential information is disclosed, users of the exchange will be less likely to continue to use the exchange for their transactions, resulting in commercial harm to the exchange. In addition, parties who have had their confidential information disclosed will be likely to sue the owners and operators of the exchange, and they may take legal action against other parties (e.g., other exchange users) involved in the unauthorized transfer of the information. This liability may exist regardless of whether the unauthorized disclosure of the proprietary material was intentional or inadvertent.

To reduce the risk of liability for disclosure of trade secrets or other proprietary content, owners and operators of commercial exchanges should implement strict systems and operational practices that will enable them to identify, manage, and secure any commercially sensitive material they process. They should also execute binding contracts with all exchange users, and those contracts should clearly define the limits of their responsibility to protect confidential material. Finally, the owners and operators of these exchanges should attempt to minimize the amount of confidential material from other parties that they collect or process.

Antitrust and competition law for trading exchanges

Antitrust law and competition law govern the ways in which businesses compete with other businesses and the ways in which businesses deal with their customers. These laws provide a framework to help promote fair competition in the marketplace. As more of the commercial marketplace moves into the electronic arena, including on-line commercial exchanges, antitrust and competition law compliance will become important issues for businesses involved in these types of e-commerce activities.

Although both antitrust and competition law are designed to protect the competitive environment in the marketplace, there are subtle distinctions between the two branches of law. Antitrust law is specifically created by statutes, while important aspects of competition law are created by courts, through the process of common law. Antitrust law most commonly carries penalties (e.g., prison terms, fines) for violations, while competition law more commonly provides for payment of monetary compensation for the actual damages caused by the illegal conduct. Although both governments and private parties enforce antitrust law and competition law by raising claims under those laws, antitrust laws are more commonly enforced by governments, while competition law claims are more commonly raised by private parties. These two categories of law are so closely related, however, that it is very common for a single lawsuit to include both antitrust and competition law claims.

A key aspect of antitrust law compliance is the ability of operators and users of on-line trading exchanges to effectively manage access to those exchanges and the information they contain. Security measures are essential to ensure that the exchanges are not intentionally or unintentionally operated in ways that are inconsistent with the requirements of antitrust and competition laws. Failure to assert that management control effectively can lead to antitrust law violations. Antitrust laws can be enforced by government authorities (e.g., in the United States they are enforced at the federal level by the Department of Justice and by the FTC, and at the state level by the 50 different state attorneys general) and by private parties who have been harmed by the violations.

One important area of antitrust law concern associated with on-line trading exchanges is that of *collusion*. Antitrust law commonly prohibits efforts by competitors to reduce competition by coordinating their business activities. An example of this type of unlawful collusion could involve two competitive businesses that agree to share customer information in order to try to drive a third competitor out of business. The policy premise behind antitrust law is that open competition in the marketplace provides the best environment for efficient matching of customer demands with market supply. If businesses coordinate the way they compete in order to reduce the overall level of competition, antitrust law presumes that the market result will be bad for consumers, in the long run.

Sharing of commercial information by businesses that compete with each other is generally a source of great concern to antitrust regulators. They are concerned about information sharing, as it can be the basis for anticompetitive collusion. When competitors share nonpublic, commercial information, they are potentially in a position to coordinate their business conduct in ways that can reduce competition. This type of activity is illegal, and it can seriously harm market competition. It is illegal if there is an intent to reduce competition (even if the attempt is unsuccessful), and it is illegal if there was no intent to reduce competition (if competition was, in fact, harmed by the actions).

Two common forms of collusion that could arise in an on-line trading exchange are price fixing and market division. Price fixing consists of an agreement (or an attempt to agree) to coordinate pricing (e.g., to set the same price for the product). When competitors attempt to coordinate the prices for their product, they are effectively attempting to limit their competition. Antitrust law prohibits attempts by competitors to coordinate their pricing decisions. Competitors have complete freedom to set prices as they see fit, based on their own independent business judgments. They are also permitted to respond to current or future pricing actions of their competitors, provided that the information as to the pricing actions of their competitors was publicly available. Competitors are not permitted to share information with their competitors regarding future pricing strategies or decisions. An on-line marketplace in which different suppliers provide the same product creates a setting in which unlawful price coordination is possible. Safeguards to prevent sharing of information on pricing strategy should, accordingly, be made an integral part of on-line exchange infrastructure and operational practices.

Another form of unlawful collusion is market division. Enterprises that compete with each other are not permitted to limit their competition by dividing up the markets that they serve. It is illegal for businesses to agree (or to attempt to agree) that they will not compete in certain markets (e.g., certain geographic regions or specific categories of customers). It is, of course, perfectly legal for each business to make independent judgments about which markets it will serve and which ones it will avoid. It is not legal, however, for businesses that compete with each other to coordinate those market entry and exit decisions with each other. On-line marketplaces create an environment in which coordination regarding market entry and exit decisions is possible, and for this reason those marketplaces must be designed and managed in ways that reduce the risk of unlawful efforts to divide up commercial markets.

From the perspective of commercial exchange owners and operators, collusion is, in large part, an information security problem. One of the most effective ways to reduce the risk of collusion is to design and operate the commercial exchanges in ways that reduce the opportunity for this type of improper information sharing. Prevention of collusion in commercial exchanges thus begins with the structure and the operational practices of those exchanges. As those structures and practices are developed and implemented, the demands of antitrust compliance should be carefully considered.

Control over access to trading exchanges and the information they contain can constitute other forms of antitrust law violations in addition to collusion to reduce market competition. For example, control over access to on-line exchanges can be used as a tool for illegal *exclusionary practices*. Exclusionary practices involve efforts by a business to harm specific competitors (or suppliers or customers) by preventing them from conducting business with specific parties. For instance, if the businesses in an industry create a trading exchange through which they plan to buy all of their component parts, any supplier that wants to sell to those businesses must participate in the trading exchange. If the owners of the trading exchange bar a specific supplier from participating, they have effectively excluded that seller from competing for the sales. That conduct by the owners of the exchange is generally considered to be an exclusionary practice, which is illegal in many jurisdictions. Another example of an exclusionary arrangement likely to be illegal is a situation in which several competitors establish an on-line marketplace through which they obtain products from suppliers. If another competitor wanted to participate in the exchange but was denied access to it, that denial of access would likely constitute an illegal exclusionary practice.

Exclusive dealing requirements are another form of antitrust law violation applicable to on-line marketplaces. Requirements that force a business to deal only with one organization can be illegal if they have an adverse effect on market competition and if there is no reasonable business justification for the exclusive dealing requirement. In the on-line marketplace environment, for instance, if the owners of a commercial exchange required that all companies that sold products to buyers using the exchange were prohibited from selling to customers through any other distribution system, that exclusivity requirement could be viewed as an unlawful exclusive-dealing requirement.

Liability for antitrust law violations is significant in many jurisdictions. In the United States, for example, antitrust law violations can result in prison terms for the individuals involved in the illegal conduct. In addition, large fines are commonly assessed against businesses found guilty of antitrust law violations. For instance so-called treble damages are often assessed for antitrust law violations in the United States. Treble damages are essentially fines in an amount equal to three times the level of the damages that the court concludes resulted from the violation of the law.

A useful summary of U.S. antitrust law concern regarding on-line trading exchanges is provided in the FTC's paper (released in October 2000), "Entering the 21st Century: Competition Policy in the World of B2B Electronic Marketplaces." The conduct of electronic marketplaces is now an area of great interest to the FTC and other antitrust regulators (particularly in the EC and in Japan). Parties involved in the development and operation of these marketplaces should make antitrust law compliance a high priority, as the penalties associated with noncompliance are substantial. The focus of antitrust compliance for owners, operators, and users of commercial exchanges should be on making access to the exchanges as open as possible (i.e., reduce the barriers to use) and on strictly controlling the sharing of nonpublic commercial information (e.g., pricing and market strategies). Effective compliance requires use of proper security technology, development of compliant exchange practices and processes, and continuous monitoring of operations to ensure effective implementation of those practices and processes.

In some jurisdictions (many states in the United States, for instance), there are competition laws that supplement antitrust laws. Competition laws are also aimed at promoting open and fair commercial competition. These laws generally permit private parties (and sometimes government authorities) to raise claims against businesses that cause damages by acting in an anticompetitive manner. For example, a business that controls information or a facility that other businesses must have access to in order to manufacture or to distribute their products can be liable for damages caused by an unreasonable denial of access to the information or facility.

Competition laws generally require that a party raising a claim under the law demonstrate that the defendant conducted business in an unreasonable and unfair manner, which resulted in quantifiable harm to the plaintiff. When the party raising the competition law claim can meet that standard, the court can order the defendant to compensate the plaintiff for damages caused by the improper conduct, and it can order the defendant to take specific actions (e.g., to change its business practices).

Most actions that constitute antitrust law violations would also be treated as violations of competition law. It is very common to see competition law claims raised at the same time as antitrust law claims. For example, a party suing for an antitrust claim of unlawful exclusionary practices would likely also raise a claim that those practices violated competition law. Competition law sometimes provides broader legal protection than antitrust law. Some anticompetitive conduct that may not qualify as an antitrust law violation could be seen as a competition law violation. For example, a party who may not have technically engaged in the antitrust law violation of illegal product tying may nevertheless be found to have acted in an unfair commercial manner resulting in harm to the plaintiff sufficient to merit a finding of competition law violation.

Effective compliance with antitrust and competition laws requires that the operators and users of on-line exchanges manage access to, and use of, the exchanges in a fair and open manner. Practices and procedures that govern exchange use must be consistent with the antitrust and competition law obligations, both in design and in operation. The operations of the exchanges must be secure enough to ensure that access denial and information sharing that could be interpreted as violations of law do not take place. Failure to provide this type of operational security can lead to significant legal liability for users, owners, and operators of commercial on-line marketplaces.

This need for effective management and security is a good reason to try to structure commercial exchanges in ways such that they are operated by parties who are independent of the users of the systems (i.e., the users of the system have no ownership interest in the exchange and have no control over the way it is operated) and that they are open to as many users as they can possibly accommodate. Commercial exchanges that are owned and operated by users of the systems raise the greatest risk of legal compliance difficulties. For example, an exchange created and operated by widget manufacturers, established to be the primary system through which those manufacturers purchased widget component parts, would raise greater antitrust concern than would arise if that same exchange were created and operated by parties not affiliated with the widget manufacturers. The level of concern would be greater, as it would be assumed that the manufacturers have economic incentives to try to operate the exchange in a manner that would give them an additional commercial advantage over other competitors not participating in the exchange, as well as advantages over suppliers and customers of the manufacturers.

As a general principle, antitrust compliance is far easier if an on-line trading exchange is owned and operated by parties independent of the users. If, however, the users of the system want to retain an ownership interest in the exchange, they should make use of separate business organizations (e.g., corporations) to manage the exchange. Using the widget industry example above, the widget manufacturers would establish a separate organization to create and operate the exchange. The manufacturers would purchase ownership interests in that separate organization, and that organization would create its own management and operational staff, independent of its owners. Although use of an independent business organization to create and operate the trading exchange will not guarantee compliance with antitrust and competition laws, the greater the independence afforded the operators of the trading exchange, the greater the protection from potential legal liability provided to the owners and the users of the exchange.

Auction fraud

On-line auctions are highly popular systems supporting a growing number of sales between individuals. The number of on-line auction transactions continues to grow at an impressive rate. As the transaction volume processed by those systems increases, the number of fraudulent transactions will also increase. On-line auction fraud poses an important legal challenge for operators and users of the on-line systems. Legal claims against fraud can be raised by both law enforcement authorities and by private parties.

Fraud is basically conduct that is intended to mislead another party, thus providing a commercial advantage to the party engaged in the fraudulent action. Fraud can consist of actual misstatements of fact or failure to disclose certain important information. Fraud can thus take place when there has been a deliberate lie or when important information has been withheld. There must be an intent to deceive or to mislead the other party, and the misleading conduct must be relevant to the decision-making process of the party who has been the victim of the fraud.

The on-line auction setting has raised issues of liability for fraudulent sale of products. Highly publicized examples of alleged fraud involving art work and other expensive goods bring the issue of on-line auction fraud to the attention of consumers and regulators. The primary issue is: To what extent does an operator of an on-line auction bear legal liability for fraudulent actions by sellers who use the auction site? Most auction site operators take the position that they should bear no liability to the buyer for these fraudulent sales, as they do not ultimately control the actions of the individuals selling through the site. Many buyers argue that the auction sites should bear some liability. Government authorities have not yet reached final conclusions on this issue, but consumer-protection regulators in many different jurisdictions are examining the issue.

There are increasing signs that at least some jurisdictions will hold on-line auction operators responsible for fraud conducted by sellers using their systems. In jurisdictions, for example, where auction operators have legal obligations to certify the authenticity of certain products, that obligation would apply whether the auction was conducted in-person or electronically. California, for instance, requires auction operators who handle the sale of sports and entertainment memorabilia to certify the authenticity of the goods sold through their systems. On-line auction operator eBay faces litigation in a California state court alleging a violation of that California state law with regard to some sports memorabilia sold using the eBay on-line auction system (*Gentry v. eBay*).

In order to reduce the chances that governments will enact regulations specifically forcing on-line auction operators to accept liability for fraudulent sales conducted using their systems, some on-line auction operators now conduct various forms of self-policing to combat the threat of fraud. For example, some operators record buyer feedback rating their satisfaction with specific sellers. Sellers who develop good reputations with their buyers are identified, while sellers who develop bad enough reputations may ultimately be banned from using the system. Some auction sites conduct periodic investigations and random audits of transactions to survey the conduct of sellers. Fees are assessed from sellers at some sites, on the theory that if a seller is required to pay to participate, the seller will be more likely to be a legitimate party. Some auction operators also permit buyers to purchase insurance that will cover at least part of their loss in the event of a failed sale. It remains unclear, however, whether these self-imposed actions by on-line auctioneers will be sufficient to prevent future government regulation of on-line auctions.

As a practical matter, on-line auction operators should assume that they will face litigation if sellers using their systems defraud buyers. The likelihood that the auction operator will be found liable for the fraudulent activity will vary from jurisdiction to jurisdiction, but on-line auctioneers should assume that they will, at some point, face litigation based on misconduct by some of their users. To act in advance to manage this risk, on-line auction site operators can take several basic steps. They should establish clear contracts (including statements defining terms of service) with all system users (both sellers and buyers) defining the limits of their liability for fraud and other failed or flawed transactions.

These contracts will not override specific statutes or regulations applicable to the user misconduct, but they can provide a source of legal rights for claims the auctioneer may choose to raise against other parties, and they may help to persuade a court that the auctioneer acted reasonably to prevent the misconduct and should thus be excused from liability. On-line auctioneers should apply the most rigorous feasible practices and procedures to try to prevent fraud and other user misconduct and to remedy the situation when misconduct occurs. Finally, on-line auction operators should consider obtaining liability insurance to help protect them in the event that serious user misconduct takes place. Although none of these advance actions will necessarily prevent fraud, they can all help to reduce the risk of fraud and to mitigate the adverse consequences if fraud occurs.

Auctioning regulated products

On-line auction popularity has led to a dramatic expansion in the variety of goods now offered for sale on those systems. Of particular interest is the rise in the number of regulated products now offered for sale through on-line auctions. Examples of these products include weapons (e.g., guns) and pharmaceutical products. The on-line auction process must be operated in a secure manner to reduce the risk that the sales transactions associated with these regulated products would violate applicable regulations.

On-line auctioneers face potential legal liability to government authorities and to private parties injured by the product if they are involved in the unauthorized sale of regulated products.

At present, most jurisdictions simply apply the rules that currently exist for the sale of these products to transactions conducted using the Internet. For instance, the rules applied to the sale of prescription medication in stores and by facsimile are also applied to the sale of medication using the Internet. One of the problems associated with this approach to regulation, however, is the fact that Internet sales commonly have a broader geographic scope than do other sales systems. The rules applicable to the on-line sale of restricted products are those of the jurisdiction in which the seller is located and those in which the buyer is located. At least two sets of regulations are applicable to each on-line sale, and because on-line transactions can involve parties in distant jurisdictions, it is sometimes difficult for the auction operator to be familiar with all of the rules applicable to each transaction.

An example of the complex nature of on-line auction compliance with applicable local laws for regulated products is provided by Yahoo!'s conviction for violation of French laws prohibiting the sale of certain hate-group materials. Goods that were barred from sale in France, but could be legally sold in the United States, were sold by a third party using the Yahoo! on-line auction system to buyers in France. Based on those transactions, a French court convicted Yahoo! of violation of the French law and ordered the company to block those auction sales. Yahoo! went to court in the United States, seeking to have enforcement of the French court's order blocked, but Yahoo! also made the business decision to block those auction sales.

The EC is currently in the process of implementing rules to address the issue of applicable law governing on-line transactions, including auctions. The Brussels Regulation, adopted by the EC, sets a standard that generally permits consumers to litigate against on-line sellers in their home country using that country's local laws. The provisions of the Rome II Regulations (which are currently being considered for adoption by the EC) generally require that, for legal disputes not involving breach of contract, the legal action can be brought in the courts and using the local laws of the country in which the act that is the basis of the lawsuit had its primary effect. The provisions of Rome II would thus enable a consumer in an EC country to raise any legal claim other than breach of contract (e.g., unfair trade practices, product liability) against a foreign on-line seller in the courts in the consumer's home country, applying the laws of the consumer's home country.

The *Yahoo!* case in France and the developing rules as to choice of forum and applicable law for consumer sales in Europe strongly suggest that on-line auction operators should be prepared to comply with the local rules regarding the sale of the products sold through their on-line systems, even if the auctioneers continue to take the position that they are not the actual sellers of the products. This apparent trend has implications for the previous discussion of fraudulent sales, as well. Although on-line auction operators may continue to argue that they are not responsible for fraud committed by sellers who use their systems, it is highly likely that local laws in many different jurisdictions will rule otherwise. We will almost certainly see an increasing number of cases, in many jurisdictions, in which local laws governing the goods and the sales transactions associated with on-line auctions will be applied to sellers and auction site operators located in foreign jurisdictions.

Application of the local laws in both the jurisdiction of the seller and that of the buyer undoubtedly complicates the operations of e-commerce retailers. Prudent on-line auction system operators should be prepared to comply with the local laws of each jurisdiction in which they conduct sales. While they may also continue to advocate the position that they should not be held responsible for sales conducted by parties who use their system and they may also continue to develop self-policing practices to manage transactions so that the risk of legal disputes decreases, on-line auctioneers should not assume that they will be able to avoid the reach of local laws in the jurisdictions in which the users of their systems are located. For operational purposes, on-line auctioneers should assume that they will be held legally responsible for the sales conducted using their systems, and they should consider insurance and other forms of protection in anticipation of the litigation in many different jurisdictions which is likely to develop.

As a practical matter, application of local law in the country of the consumer places a significant burden on the seller. Sellers must be aware of the local rules in a great number of jurisdictions if they engage in on-line transactions, and they must develop practices and procedures that ensure compliance with those local rules. This burden is one that has long been managed by large retailers that have had global operations for a substantial period of time. It is a great challenge, however, for small and midsize businesses that have traditionally been local in scope, but have recently expanded to international operations thanks to the development of the Internet. Application of local laws thus appears to provide a competitive advantage to larger retailers by requiring compliance with many different legal obligations in many different jurisdictions. It is an open issue whether this is the result that governments actually intended.

Intellectual property in on-line auctions

On-line auctions have also become venues where individuals offer a variety of goods protected by intellectual property laws for sale. Computer software, music recordings, video content, and images are all examples of intellectual property that is now widely available for sale on on-line auctions sites. These sales raise legal issues associated with the protection of intellectual property rights. Copyright owners express concern over the apparently increasing amount of copyright protected material being sold through on-line auctions. They are concerned about both the auction sale of pirated material and the sale of copies of material that may have been legitimately purchased by a consumer, for although a consumer who buys a copyrighted product (e.g., software) owns that single copy and has limited rights to make duplicates of the product for his or her own personal use, that legitimate buyer does not have the right to sell copies of the product. Some copyright owners fear that on-line auctions create a setting in which it is extremely difficult to enforce effectively copyright ownership rights, and they want to hold on-line auctioneers legally responsible for infringement.

The issue is the extent to which the on-line auctioneers are legally responsible for the intellectual property law violations of the users of their systems. When a seller auctions a music file that has been pirated, for example, is the auctioneer liable for contributory infringement of the copyright for the music, under the theory that the auction-site operator facilitated and enabled the unlicensed distribution of the material? This issue is, at present, unresolved. Copyright owners (e.g., the music recording industry) are urging on-line auction operators to take aggressive action to prevent unauthorized distribution of protected material, but the extent of auctioneer legal responsibility for infringement by its users remains uncertain.

If current trends in intellectual property protection applied to other on-line distribution systems provide a meaningful guide, it appears that on-line auctioneers will face at least some liability for auction sales involving copyrighted material in the future. Napster and other facilitators of peer-to-peer content distribution have been held accountable for unlicensed circulation of copyrighted material by their users. Just as these distributed-computing systems now face liability even though they are not the actual provider of the unlicensed material, so too is it likely that on-line auctioneers will be held liable, at least in part, for copyright infringement that takes place using their auction systems.

Certain jurisdictions, the United States for one, now provide some measure of legal protection for ISPs in regard to content generated by other parties, which they make available. For example, the Digital Millennium Copyright Act in the United States provides ISPs with protection from liability for copyright infringing material that is distributed by others using their networks, provided that the ISPs implement reasonable practices and procedures to prevent such infringement and that they take reasonable steps to remove infringing material promptly when it is identified. Napster and other facilitators of peer-to-peer distribution have tried to argue that they should qualify for protection from liability under the ISP exemption of the act, but to date they have not been afforded that protection. On-line auctioneers will likely find themselves in a similar position.

On-line auctioneers should take several basic steps to reduce the potential risk of liability resulting from copyright infringement by users of their systems. They should structure their contracts with users (including their terms of service) to require that all users of the systems avoid misuse of the intellectual property of other parties. Those contracts should also indicate that failure to comply with this requirement could result in immediate termination of the right to use the system. Those agreements should also specifically define the process to be used to advise the auctioneer of suspected instances of intellectual property misuse, and they should indicate that the auctioneer has the right to stop transactions and to remove content if the auctioneer concludes that intellectual property rights are being violated. Operators of on-line auctions may also want to obtain liability insurance to protect themselves in the event of intellectual property infringement by their users.

Property rights for information

On-line auctions and commercial trading exchanges rely on information accessibility and sharing to perform their functions. In that environment, the issue of ownership and control over information is a critical one. Some commercial parties now assert property ownership claims over information and data files that they develop. The extent to which those claims are enforceable has not yet been fully defined; however, the existence of property rights claims over commercial information must be addressed by operators and users of on-line auction and commercial-trading exchange systems. Recent litigation in the United States (e.g., *eBay v. Bidder's Edge* and *House of Blues v. Streambox*) illustrates the fact that property rights asserted over on-line content can be a significant factor for e-commerce operations.

The property issue should be examined from two perspectives. Developers and operators of electronic commerce systems should be aware of their potential right to assert control over their data content and over the data files that create their on-line presence (e.g., Web sites). Developers and system operators must also, however, be aware that other parties may assert the same rights over their data content. Application of traditional property rights to data collections and data files thus presents a double-edged sword. It provides an additional legal right that the owners of the digital property may try to enforce to exercise control over the use of the material. Yet it also provides an additional potential legal liability that may be enforced against an e-commerce operator when that operator attempts to make use of digital property owned by other parties. Operators and users of electronic commerce systems are owners of their own digital property and users of the digital property of others.

A party who uses another's digital property without appropriate authorization faces at least three forms of potential legal liability: a property law claim such as conversion, liability for the misuse of the personal property of another party; for breach of contract, if a binding contract was in place, prohibiting use of the digital property in question; and liability for an unfair business practice, particularly if the unauthorized use of the digital property resulted in commercial harm to the owner of the property.

It is also possible that an owner of digital property could face legal liability if his or her actions to control access to that property are unreasonable. For example, if an owner of property denies access to the property with the intent to drive another enterprise out of business, the denial of access could be viewed as an unfair business practice. Similarly, if the property in question is some form of *essential facility* (an antitrust law concept involving property that must be accessed by all the companies in an industry if they are to compete effectively), unreasonable denial of access to that property could constitute an antitrust law violation.

If you are a digital property owner and you want to limit access to your property, the best approach is to apply all active protective measures (e.g., technologies and systems) that you can to restrict access. For example, use of registration systems and robot exclusion files (to help limit access by automated search programs) are common practices and can be effective. Reliance on these self-help measures to control access to the digital property is likely to yield greater success than will efforts to apply legal remedies through litigation and other means.

If you are a user of the digital property of others you should comply with all of the reasonable restrictions placed on use of the property by the owners of the property. The law appears to be accepting at least some principles of property ownership applied to digital content, even though the exact scope of that acceptance remains unclear. In that environment, failure to comply with reasonable controls on digital property use is likely to result in conflicts that can lead to legal penalties.

Rights and duties of a provider of outsourced services

As more information technology systems and functions supporting electronic commerce activities are provided by outside contractors on an outsourced basis, issues associated with sharing and enforcement of legal rights and obligations become more complex. Enterprises commonly make use of contractors to plan, implement, operate, and maintain their e-commerce activities. These outsourced functions cover a wide range of activities supporting electronic commerce transactions. They may include software development, Web site development or hosting, application service provision, payment processing, database management, or any of a variety of other e-commerce support activities. The rights and duties associated with security of the contractor providing the outsourced services and of the client must be clearly established and enforced. Each time that a contractor is introduced into the e-commerce system, there is another party who can both raise legal claims and be the target of those claims. Use of contractors thus inevitably complicates the issue of legal compliance.

One of the key legal issues associated with the security of outsourced services is that of protection of trade secrets and other confidential material. Many e-commerce transactions involve the transfer or use of proprietary material that may involve trade secrets or private information. As more e-commerce functions are shifted to contractors, those contractors will commonly be involved in the collection and processing of this protected information. Client and contractor should expressly agree regarding who will be responsible for the security of trade secrets, personal information, and other confidential or proprietary content processed by the system. That understanding should always be memorialized in a clearly written contract.

Another important issue is the appropriate sharing of liability in the event of a failure of the system. In effect, the contractor and the client must reach some accommodation as to risk of loss associated with security breaches. This process of sharing should address situations in which the system failure results in harm to the client and when it leads to harm to a third party. The contractor and client should discuss this issue in advance and should reach a precise agreement regarding who will bear liability or the extent to which that liability will be shared. The terms of this agreement should be clearly expressed in a written contract.

A common principle applied to the issue of liability sharing is to try to place legal liability on the party who has the most control over the actions that will lead to the liability. Many parties conclude that fairness dictates that a party should not be held responsible for liability that results from actions that it did not control. Instead, they often conclude that liability should be borne by the party who had the best opportunity to prevent the problem. Using this approach, the parties might agree, for example, that liability for damages to a third party caused by a breach of security for a database containing credit card account numbers would fall to the contractor who was responsible for creating and maintaining that database.

Another commonly used approach to liability sharing is the use of indemnification and insurance. The parties may supplement their agreement as to how liability will be apportioned by agreeing that one party will compensate the other party in the event the compensated party suffers costs as a result of some failure on the part of the party who has the obligation to pay the compensation. This process of compensation is often referred to as a right of indemnification, and it permits the parties to share costs associated with liabilities. In some instances, contractor and client may also agree that one or both of them will obtain insurance to provide protection in the event of legal liability. Use of insurance requirements provides another effective liability sharing mechanism.

Use of contractors to provide outsourced services in the digital marketplace is made necessary by the technical and commercial demands of that marketplace. Effective security management requires that all parties in those outsourcing relationships understand and perform fully their duties. Development of those relationships merits careful consideration and thorough planning. Failure to invest that caution in advance can result in loss of legal rights and increased risk of third-party liability for both contractors and clients.

Appendix 8A: Legal guidelines for on-line auctions

Operators of on-line auction sites face significant potential legal liability. To minimize the risk of claims, and to enhance the prospects of successful defense if claims are raised, auction site operators should consider the following basic guidelines.

Establish enforceable contracts

Written contract terms should be established with all users (buyers and sellers) of the system. Click-through agreements are acceptable, and they should be used in conjunction with clearly written terms of service.

Specify limits of liability

Contracts with users should clearly and thoroughly explain limits of liability. The auction operators should expressly indicate that they are not responsible for misconduct by any users of the system. The effectiveness of this type of disclaimer and liability limitation is, of course, uncertain in many jurisdictions, but it should be included in the agreements nonetheless.

Right to stop transactions and to ban users

Auctioneers should retain the right to stop any transaction they believe to be illegal or otherwise inappropriate. They should also expressly reserve the right to terminate usage privileges for any system user in the event of system misuse.

Communications with users

Clearly described procedures to enable system users to communicate with the auctioneer should be established. The auctioneer should invite questions, comments, and complaints from all users, and should provide a specific contact person (and contact information) for communications from users.

Protections for users

As a business judgment, on-line auctioneers should take steps to protect system users. Effective potential measures include development of seller rating systems, assistance in payment processing for transactions, and insurance for system users.

Multijurisdictional compliance

Auctioneers should be prepared to comply with local laws in all jurisdictions in which their buyers and sellers reside. Although this may be a challenging undertaking, it is likely to reduce substantially the risk of future liability.

Appendix 8B: Legal principles for on-line commercial exchanges

Internet-based commercial exchanges present challenging issues of legal compliance. Operators and users of those on-line marketplaces should focus on legal compliance as they design and manage the operations of those marketplaces. The following basic legal principles form an effective foundation for compliance practices.

Establish written contracts governing use

Exchange should establish binding contracts with all users. Click-through agreements with clear and complete descriptions of terms of service are appropriate.

Antitrust and competition law compliance

Exchanges should be designed and operated in ways that comply with laws protecting fair commercial competition. When feasible, exchanges should be owned and operated by neutral parties (i.e., parties who are not buyers or suppliers in the exchange). Exchanges should promote open access for all interested and qualified parties. Information sharing among business competitors should be carefully controlled.

Information protection

Exchange operations should protect trade secrets, personal information, and other forms of confidential or proprietary material effectively. Management of sensitive information should be a high priority for exchange operators.

Communications with users

Exchange operators should establish clear means of communications with all exchange users. Questions, comments, and complaints should be invited and promptly addressed. Specific instructions as to how to communicate with the exchange operators (e.g., specific contact information) should be provided to all users and should be kept current at all times.

Multijurisdictional compliance

Exchange operators should be prepared to comply with the local laws of all jurisdictions in which the exchange conducts business (e.g., the jurisdictions in which its users are located).

Appendix 8C: Managing legal compliance for outsourced services

Many different information technology systems and operations are commonly performed by outside contractors. This popular trend is becoming more prevalent in the context of support for electronic commerce operations. To provide for effective management of legal compliance as to security issues by both the contractor and the client, the following fundamental topics should be addressed.

Select parties carefully

Parties involved with e-commerce activities should select their contractors carefully, largely basing that decision on the verifiable reputations of the contractors. The most effective step toward reducing the risk of future legal liability is to work with well-established and highly qualified providers of outsourced services.

Create a written contract

The terms of the understanding between client and contractor for outsourced service should be defined in a binding, written contract.

Defining responsibility and authority

Contractor and client should reach agreement on the distribution of responsibility with regard to system and transaction security. Their contract should clearly define the specific responsibilities borne by each party with regard to security. It should also establish and describe the authority that each party will possess with respect to actions necessary to meet security requirements and actions taken in response to security threats and failures.

Liability sharing

The parties should expressly agree as to how they will share liability in the event of security failures. The issue of liability should address circumstances when a claim is available to either the contractor or the client and instances when the liability involved is a claim raised by a third party.

Monitor performance

Performance by both client and contractor should be monitored consistently during the term of the agreement. Systems and standards to support effective monitoring should be agreed upon in advance by the parties and should be defined in the contract.

9

Conclusion: Future Trends and Challenges

The digital marketplace continues to evolve. In the future, it will be shaped by many different trends and forces. That evolution will have significant continuing impact on security concerns and it will dramatically affect the enforcement of legal rights and liabilities associated with electronic commerce security. This chapter highlights some of the more important forces that are likely to play a role in shaping the digital marketplace of tomorrow, and it underscores some of the important legal issues related to security in that marketplace.

The ubiquitous network

Ever smaller, more powerful, and less expensive computing devices integrated into networks that make greater use of wireless communications technology will create a virtually omnipresent computer network, the ubiquitous network. The ubiquitous network brings tremendous potential for economic, political, and social benefits. It will also have major implications for system security and the enforcement of rights and obligations associated with that security.

As computers and network access become pervasive, the security challenges traditionally faced by classic computer networks will be confronted by individuals and organizations that do not have extensive security management experience. When consumer electronics equipment, household appliances, automobiles, and most of the other devices we consider to be basic components of everyday life become part of the public computer network, the majority of the public will have a direct and obvious stake in the security of that network. In that setting, the number of parties who have legal rights to network security for their content and transactions is very large, indeed.

At the same time that the number of parties enforcing their legal rights increases, the ubiquitous network will also dramatically increase the number of parties with obligations to protect content and transaction security. More parties will be directly involved in the construction, operation, and maintenance of this network. In the world of the ubiquitous network, the parties facing legal responsibilities for protection of security will include many players who currently face no such requirements. For example, when household-appliance manufacturers commonly build and sell Web-enabled appliances, those manufacturers will effectively become a part of the computer security sector. Consumers of their products will expect that those appliances will be compatible with system-security standards and requirements. If those products fail to function in a manner consistent with security for the system, the manufacturers will be among the first targets of legal actions raised by consumers and government authorities.

In the world of the ubiquitous network, security threats also proliferate. As more content and transactions are processed by the network, incentives for unauthorized activities will increase. The breadth of the overall public network also provides more access points and more tools for those with intentions to misuse the equipment. The ubiquitous network thus increases both the motives for malfeasance and the means to breach the security of the network. In addition, the ubiquitous network makes the potential adverse effects of security breaches greater than they are today. When the network is fully integrated into more functions used by more people in more parts of the world, breaches of network security carry far greater potential for harm than they do with a more limited network.

The ubiquitous network makes mobile commerce practical. Mobile commerce brings great potential to serve the demands of businesses and consumers. Early experience with mobile commerce also suggests, however, that the security requirements for mobile commerce are significant. Mobile access to the digital marketplace will likely become increasingly popular in the near future and in many different parts of the world. This mobile, digital marketplace will also challenge those responsible for security to develop effective measures to protect the content and transactions supported by that marketplace.

Diversity of content and applications

The array of content delivered by the Internet and applications supported by it will continue to expand dramatically. In that environment, huge numbers of users will make use of the global public information network to deliver diverse content and to enable a virtually limitless variety of transactions and activities. The public information network will thus be called upon to manage the diverse rights and duties associated with a dazzling range of content and relationships among network users. Different content and different relationships will require different levels of security, all to be supported by one public system. This is no small challenge.

One area of increasing attention affecting diverse on-line content is the issue of digital-rights management. Diverse material (e.g., software, video, audio, images, text) accessible through the Internet requires effective systems and practices to permit adequate control of ownership. Efforts to facilitate digital-rights management will be a major element of on-line security initiatives. Traditional intellectual property rights will continue to be a main component of digital-rights security, but they will be supplemented by other legal principles. Contract and commercial transactions law, for example, will increasingly be applied to digital-content management. We will also see property-law theories applied by content owners as part of their effort to manage digital rights. Copyright, trademark, and other forms of intellectual property law have been at the core of digital-rights management to date, but in the future those legal rights will likely be supplemented by commercial law and property law to provide a more effective basis for rights management efforts.

Distributed computing

Future commercial use of the Internet will be characterized by increasing reliance on distributed-computing systems. Peer-to-peer and other formats that facilitate greater user control over content sharing will gain in popularity. As more content control shifts to system users, management of that content becomes a greater challenge. Security for that content is more difficult to maintain. Legal rights and duties are tougher to enforce. Increased use of distributed-computing systems poses one of the greatest challenges for effective enforcement of legal rights in the future.

In a distributed-computing environment, system users must take a more active role in digital-security management. In a network with a hierarchical structure, there are content access "choke points," where content-security controls can be more readily applied. Distributed systems are more diverse and can reduce the number of content control points. This diversity of access and control is one of the strengths of the distributed structure. It is also one of the big challenges for security. In this distributed-content environment, individual users face greater security threats and bear greater security responsibility, as the ability of intermediaries to perform content-security functions is diminished by the content-sharing structure. Some have argued that these systems empower the end user. Along with that power, however, come threats and obligations. Users of distributed-content systems must make sure that they recognize the legal risks associated with those systems and that they are prepared to manage those risks appropriately.

Open-source content

The current trend toward more widespread acceptance of the open-source model for software and other content accessible on-line will continue. Over time, more of the goods distributed through the digital marketplace will be provided using some form of open-source model. Management of that content will be more difficult. Protection of proprietary material in a setting of widespread open-source distribution will be an important challenge likely to require nontraditional views as to intellectual property security and control.

Security for open-source content will primarily involve the following legal concerns. Developers using open-source material must make sure that their use of that material complies with the terms of the open-source license, and some parties must be empowered with the authority to enforce the open-source license obligations. Even though the open-source model is premised upon widespread access to source code, that model continues to involve an intellectual property license, and the terms of that license must be effectively enforced.

Open-source developers must also manage the issue of product fragmentation. As more modifications and applications are created, it is important to provide a means through which the various versions of the open-source product will be compatible with each other. Some degree of fragmentation is virtually inevitable, given the open-source development dynamic, but that fragmentation should not be allowed to create incompatible versions of the basic product. Security for open-source material thus involves efforts to ensure compliance with the licensing requirements associated with open-source material provided by others and actions designed to enforce your own open-source terms in order to reduce the risk of product incompatibility arising from excessive fragmentation of those products.

On-line communities for collaborative commerce

More economic and commercial relationships will migrate to the Internet. Individuals and organizations will develop collaborative electronic relationships, and those relationships will shift over time. In this environment, on-line communities will be established to facilitate collaborative commerce, where an ever-evolving set of suppliers and consumers find ways to work together for mutual gain. Two key attributes define on-line collaborative commerce: diversity of participants and continuing evolution of relationships among participants. Those two characteristics have a profound impact on rights and obligations associated with community security.

Collaborative commerce in the digital marketplace will require that the computer and information systems that support commerce must effectively enable both sharing, and restricting, of access to information. Collaboration consistent with economic benefit and protection of legal rights requires highly sophisticated management of e-commerce transactions. To date, there has been more attention paid to the development of information-sharing systems and processes, than has been paid to true information management in these emerging collaborative digital environments. In the future, developers, operators, and users of collaborative commercial communities will recognize that, in order to gain economic value from those communities, they must structure the communities so that information is truly managed. Only in those communities in which there is a proper balance of content sharing and content controls will the economic value of the collaboration be realized in a manner consistent with legal obligations.

The content and transaction management systems for these collaborative commercial communities must also accommodate continuous change. The participants in those communities will change frequently, as will the content and the policies applied to access and use of the content. Despite these continuous changes, the system must effectively regulate access to and use of the community in order to preserve the required level of security. More widespread participation in these digital collaborative communities will thus create a major operational challenge for the electronic marketplace. Failure to meet that challenge will lead to substantial legal liability.

Increasing regulation and multijurisdictional compliance

As the scope of electronic commerce expands, pressure to increase regulation of that activity will also increase. That pressure will likely be characterized by two important factors. One factor will involve demand for new rules to be applied specifically to the e-commerce environment. Care should be exercised to make sure that any regulations devised for the digital marketplace are consistent with those applied to traditional commerce. The second factor will involve development of regulatory oversight in many different legal jurisdictions. The trend toward greater regulation of the digital marketplace by many different governments around the world will have a profound impact on rights and duties associated with the security of that marketplace.

Electronic commerce security will be one of the most popular targets for future regulation in virtually all jurisdictions. We will almost certainly see governments compete with each other in a race to show that they are the most responsive when it comes to reducing digital commerce security threats. Unfortunately, much of this regulation will likely be misguided and a significant amount of it will be improperly enforced. Do not assume that security regulation will develop efficiently or that it will be particularly effective, at least in its early stages, at reducing threats to the digital marketplace.

There will also likely be some disparity between regulation aimed at digital-security concerns of individual consumers and those directed toward business-to-business commerce. Expect greater variety and more political tone for regulations applicable to consumer security in the digital marketplace than that for regulations affecting electronic commerce transactions among businesses. Governments are likely to enact laws to address e-commerce security concerns of individual consumers, while relying instead on negotiated arrangements (e.g., contracts) and private legal actions to establish and enforce legal rights associated with digital commerce security among businesses.

There will also likely be economic pressure on governments to achieve at least a basic level of consistency between their digital marketplace rules and those of other governments. For example, if one government enacts e-commerce security regulations imposing a commercial burden on e-commerce operations that is substantially greater than that imposed by other governments, the government taking the more stringent action may place its jurisdiction at a competitive disadvantage relative to the other jurisdictions. Most governments will want to avoid this type of disparity that places their jurisdiction at a competitive disadvantage. The desire to avoid this situation will likely lead many of the jurisdictions that have significant digital marketplace aspirations to establish e-commerce rules that are consistent with those of other jurisdictions.

Self-defense in the digital marketplace

In addition to increased legal and regulatory oversight in the digital marketplace, participants in that marketplace will see the need for greater reliance on their own actions to reduce the risk of security breaches. The inherent problem with legal remedies is that they can only be applied after the fact, after the harm from a failure of security has been suffered. Legal remedies generally provide for compensation to those who have been harmed or sanctions against those who acted illegally. Laws and regulations prevent security compromise only to the extent that they deter deliberate misconduct, through their threat of active enforcement and severe penalties, or they require protective measures that actually serve as effective barriers to the security breaches. In contrast to legal remedies, promotion of self-help measures to protect digital security can help to prevent security failures.

Already today, we see that commercial enterprises are investing more resources in security for electronic commercial activities. Those businesses now recognize that the adverse consequences associated with threats to security are significant and that the ability of legal remedies to block those threats or to compensate adequately for the damage caused by security breaches is severely limited. This trend is likely to continue for the foreseeable future, as more commercial participants in the digital marketplace discover that the best approach to security protection is prevention.

Individual consumers engaged in electronic commerce will also increasingly appreciate the importance of on-line security. With this recognition will come greater willingness to invest time and resources in efforts to reduce their personal security risks. Knowledge will provide the strongest weapon to enable consumers to protect their security in the digital marketplace. Knowledge of the threats they face and knowledge of the methods to respond to those threats will have the greatest positive impact on individual e-commerce customers.

Security as a management issue

Perhaps the most important trend for security in the digital marketplace is the increasing willingness to address electronic security for what it is, a management issue. Protection of e-commerce security requires effective management of several fundamental resources: technology, people, and information. Failure to manage any one of those resources effectively will threaten digital security. Security in the electronic commerce environment is thus a resource management challenge, not solely a technology or legal problem. New technologies or more aggressive laws will not make the digital marketplace more secure, unless they are coupled with more effective management of all of the key resources that drive e-commerce. We will likely find that the digital marketplace participants who prove to be the most adept at protecting security will be the ones that are the best managed.

Security as a competitive advantage

In the digital marketplace of tomorrow, effective security will almost certainly be an important source of competitive advantage. Enterprises that provide secure environments for electronic commerce transactions and relationships will be the most popular businesses in the eyes of all of the major stakeholders: customers, business partners, investors, and regulators. We will most likely find that attention to digital security not only reduces the risk of legal liability, but also enhances commercial prospects, providing competitive benefit.

Final thoughts

To effectively protect your legal rights and to minimize your risk of legal liability, digital commerce security must be a high priority in your business planning. Security is not solely a technical or a legal issue, but is instead a resource and risk management issue. It is your choice how you decide to allocate your technology, people, and capital resources to balance the risks and rewards associated with digital commerce security. As you make those resource allocation decisions, there are two basic principles worth highlighting.

First, the range of enforceable legal requirements applicable to the digital marketplace is currently complex and is virtually certain to be substantially more complex in the future. The digital marketplace is a major part of the global economy. As the number of participants in that marketplace increases, and the variety of their transactions and relationships expands, more legal requirements will be applied to the marketplace. As the value of the digital marketplace grows, the number of threats to the marketplace will proliferate and the potential adverse consequences of those security threats will increase. In that setting, enforcement of your organization's rights and compliance with its legal duties, in the context of digital security, will be a major challenge. Management of digital security in the e-commerce marketplace of tomorrow will place far greater demands on all participants in that marketplace than they encounter today.

Second, compliance with the diverse set of legal requirements for security in the electronic commerce environment is both a duty and an investment that can lead to future benefit. To date, many have focused on the legal duties associated with digital security, highlighting the significant potential legal and economic liability if e-commerce participants fail to provide adequate security. Tomorrow, more digital marketplace participants will recognize that protection of security is an investment that provides value, independent of its usefulness in reducing legal liability risks.

Marketplace participants who provide a secure e-commerce environment and comply with all applicable legal obligations for security gain competitive advantage. They gain this advantage by being more attractive to customers, investors, business partners, and regulators. In addition, by respecting the rules and the rights of others, those market participants contribute to the creation of a commercial setting in which other parties are more likely to respect rights, and all participants in the digital marketplace will benefit, in the long-term, from such an environment. Attention to digital commerce security is necessary from a legal perspective and it is good business.

Selected Bibliography

1267623 Ontario, Inc. v. Nexx Online, Case No. C20546/99, Ontario Cup. Sup. Ct. 1999.

"Advisory Panel of the U.S. Congress to Assess Domestic Response Capabilities for Terrorism Involving Weapons of Mass Destruction," *Second Annual Report*, 2000, at http://www.rand.org/terror2.pdf.

ACLU v. Reno, 117 S. Ct. 2329 (1997).

ACLU v. Reno (Reno II), 31 F. Supp. 2d 473 (E.D. Pa. 1999).

Amazon.com v. Barnesandnoble.com, Case No. C99-1695P (E.D. Wash. 1999).

American Civil Liberties Union (ACLU), *Project Echelon Watch*, 2001, at http://www.aclu.org/echelonwatch.

American Guarantee & Liability Ins. Co. v. Ingram Micro, Inc., 2000 U.S. Dist. LEXIS 7299 (D. Ariz. 2000).

AT&T Corp. v. City of Portland and Multnomah County, Case No. CV 99-65-PA (U.S. Dist. Ct. Or. 1999), *rev'd*, 216 F.3d 871 (9th Cir. 2000).

Australian Crimes Act, Part VIA.

Austrian Privacy Act, Sect. 10.

Berkman, E., "Stamps of Approval," *CIO*, Mar. 1, 2001.

Bernstein v. United States Dept. of Justice, Case No. 97-16686 (9th Cir. 1999).

Blumenthal v. Drudge, 992 F. Supp. (D.D.C. 1998).

Bourke v. Nissan Motor Corp., No. BO68705 (Cal. Ct. App. 1993).

British Telecom v. Prodigy (S.D.N.Y., filed Dec. 13, 2000).

Bunyan, T., "EU Governments to Give Law Enforcement Agencies Access to All Communications Data," *Statewatch*, May 16, 2001, at http://www.statewatch.org.

Canavan, J. E., *The Fundamentals of Network Security*, Norwood, MA: Artech House, 2000.

Carlton, J., and P. W. Tam, "Online Auctioneers Face Growing Fraud Problem," *Wall St. Journal*, May 12, 2000, p. B6.

Cello Holdings v. Lawrence-Dahl, 89 F. Supp. 2d 464 (S.D.N.Y. 2000).

Church of Scientology v. Dataweb, Case No. 96-1048, Dist. Ct. of the Hague, Holland, 1999.

CompuServe, Inc. v. Cyber Promotions, Inc., 962 F. Supp., 1015 (S.D. Ohio 1997).

Computer Crimes Act, Malaysia, 1997.

Computer Fraud and Abuse Act of the United States, 18 U.S.C.A., Chap. 47, Sect. 1030.

Computer Law of Israel, Sect. 4, 1995.

Computer Misuse Act of Singapore, Chap. 50, Sect. 3.

Computer Misuse Act of the United Kingdom, Chap. 18, 1990.

Cotrone v. RealNetworks, No. 00CV 2629, (N.D. Ill. 2000).

Council of Europe, *Brussels Regulation*, 2000, at http://europa.eu.int.

Council of Europe, *Draft Convention on Cyber-Crime*, 2000, at http://conventions.coe.int/treaty/en/projets/cybercrime.htm.

Council of Europe, *Framework for Electronic Signatures*, 2000, at http://europa.eu.int.

Council of Europe, *Intellectual Property in the Information Society Directive*, 2000, at http://europa.eu.int.

Council of Europe, *Rome II Directive*, 2000, at http://europa.eu.int.

CRIM. CODE of Canada, § 342.1.

CRIM. CODE of Greece, art. 370C, § 2.

CRIM. CODE of Luxembourg, § VI, art. 509-1.

CRIM. CODE of The Netherlands, art. 138(a).

CRIM. CODE on Computer Crime, Belgium, art. 550(b), 2000.

Criminal Damage Act of Ireland, Sect. 5, 1991.

Criminal Information Law of Portugal, Chap. 1, Art. 7.

Dash, J., "Security Top Concern as Health Care Regs Loom," *Computerworld*, Feb. 12, 2001, p. 73.

Data Act of Sweden, Sect. 21, 1973.

Delta, G. B. and J. H. Matsuura, *Law of the Internet*, New York: Aspen Law & Business, 1998, updated 2001.

Digital Signature Act, Malaysia, 1997.

Digital Signature Act, South Korea, 1999.

Digital Signature Law, Germany, 1997, at http://www.iid.de/jukdg/sigve.html.

Digital Signature Legislation, Argentina, 1998, at http://www.sfp.gov.ar/decree427.html.

Dorer v. Arrel, 60 F. Supp. 2d 258 (E.D. Va. 2000).

eBay, Inc. v. Bidder's Edge, Inc., 100 F. Supp. 2d 173 (N.D. Cal. 2000).

"eBay Suspends the Seller of Painting in Its Auction," *Wall St. Journal,* May 11, 2000, p. B5.

Electronic Commerce Act, South Korea, 1999.

Electronic Communications Act, United Kingdom, 2000.

Electronic Communications Privacy Act of the United States, 18 U.S.C.A., Sects. 2701-2710.

Electronic Exchanges & Electronic Commerce Law, Tunisia, 2000.

Electronic Signatures in Global and National Commerce Act (E-Sign), United States, 2000, Public Law 106-229, at http://thomas.loc.gov.

Electronic Signatures Law, Poland, 2001.

Electronic Transactions Act, Bermuda, 1999.

Electronic Transaction Regulations, Singapore, 1999, at http://www.cca.gov.sg.

Export Control Regulations, United States, 2000, at http:www.bxa.doc.gov/EncryptionRuleOct2K.pdf.

Fausett, B. A., "Linking Legalities," *Webtechniques,* Feb. 2001, p. 18.

Financial Services Modernization Act, United States, 2000, at http://thomas.loc.gov.

Flanagan v. Epson America, Inc., No. BC007036 (Cal. Super. Ct. 1991).

Foster, E., "Avoid Getting Burned by Terms of Software Licenses in the Age of UCITA," *InfoWorld,* July 10, 2000, p. 89.

"FTC Assesses First Fines for Violating Kids On-Line Privacy Law," *Computerworld,* April 19, 2001, at http://computerworld.com/nlt.

FTC, *Entering the 21st Century: Competition Policy in the World of B2B Electronic Marketplaces,* 2000.

FTC v. Netscape Communications Corp., Case No. CV-00-00026-(Misc.) MHP (N.D. Cal. 2000).

FTC v. Toysmart.com LLC, Civ. 00-11341-RGS (D. Mass. filed 2000).

Gellman, R., "The Maze of New Health Privacy Rules," *DM News*, Jan. 29, 2001, p. 14.

Gentry v. eBay, Case No. 746980 (San Diego Sup. Ct., Cal, filed 2000).

Genusa, A., "Conspiracy of Silence," *CIO*, March 1, 2001, p. 93.

Goodman, P. S., "911 Still Can't Pinpoint Calls From Cell Phones," *Washington Post*, April 8, 2001, p. A1.

Greene, T., "Forum Warns of Hidden DDoS Legal Liability," *Network World*, Oct. 2, 2000, p. 16.

Gross, G., "Microsoft's Passport Service: No Marylanders Allowed?" *Newsforge.com*, April 26, 2001, at http://www.newsforge.com.

Guy Laroche v. G.L. Bullentine Board, Tribunal de Grand Instance de Nanterre, 2000.

Harrison, A., "New Wave of Threats Against Your Data," *Business 2.0*, Mar. 6, 2001, p. 50.

Hassler, V., *Security Fundamentals for E-Commerce*, Norwood, MA: Artech House, 2000.

Health Insurance Portability & Accountability Act (U.S.), Pub. Law 104-191, 1996, at http://www.hipaadvisory.com.

Illinois Institute of Technology Research Institute, *Study of Carnivore Internet Surveillance System Conducted for the FBI*, 2000, at http://www.usdoj.gov/carnivore_draft_1.pdf.

Information Technology Act, India, 2000.

Insituform Technologies, Inc. v. National Envirotech Group, LLC, Case No. 97-2064 (E.D. La. 1997).

Intel Corp. v. Hamidi, 1999 WL 450944.

Intellectual Reserve, Inc. v. Utah Lighthouse Ministries, Inc., Case No. 2:99-CV-808C (D.C. Utah 1999).

John Doe aka Aquacool_2000 v. Yahoo!, Inc., Case No. 2:00 CV 04993 (C.D. Cal. 2000).

John Tesh v. Celebsites, Inc., Case No. 00-00603-ABC (RZX) (C.D. Cal. 2000).

Johnson, D. G., "Employee Monitoring: Drawing the Line," *Beyond Computing*, Nov./Dec. 2000, p. 16.

Julia Fiona Roberts v. Russell Boyd (WIPO Domain Name Arbitration), Case No. D2000-0210 (2000).

Law on Automated Data Processing, Chile, 2000.

Law on Data Messages & Electronic Signatures, Venezuela, 2001.

Lemos, R., "Digital Signatures a Threat to Privacy?" *ZDNet News*, April 7, 2000, at http://www.zdnet.com/zdnn/stories/news.

Levitt, J., "Wireless Devices Present New Security Challenges," *InformationWeek*, Oct. 23, 2000, p. 120.

Lorek, L., "Share the Risk," *Interactive Week*, Feb. 19, 2001, p. 16.

Mangalindan, M., "Alleged Drug Sale on eBay Raises Liability Issue," *Wall Street Journal*, May 30, 2000, p. B18.

Mangalindan, M., and K. Delaney, "Yahoo! Ordered to Bar the French from Nazi Items," *Wall Street Journal*, Nov. 21, 2000, p. B1.

Matsuura, J. H., *A Manager's Guide to the Law And Economics of Data Networks*, Norwood, MA: Artech House, 2000.

McGinley, L., S. Lueck, and J. VandeHei, "Patient-Data Rules to Go into Effect," *Wall Street Journal*, April 12, 2001, p. A18.

McLaren v. Microsoft Corp., 1999 Tex. App. LEXIS 4103 (Tex. Ct. App. 1999).

Messmer, E., "Federal Net Privacy Mandate Riles Healthcare Industry," *Network World*, Feb. 12, 2001, p. 1.

Miles, L., "Rome II, Internet 0," *eCFO*, Spring 2001, p. 13.

Mitchener, B., "Group Seeks PC Royalties from Fujitsu Siemens," *Wall Street Journal*, April 9, 2001, p. A22.

Mitchener, B., "Microsoft Plans to Sign Accord on Data Privacy with the EU," *Wall Street Journal*, May 16, 2001, p. A14.

Nagel, K. D., and G. L. Gray, *Electronic Commerce Assurance Services*, New York: Harcourt Brace, 1999.

National Federation of the Blind v. AOL, No. 99CV 12303 EFH (D. Mass. 2000).

Network Solutions, Inc. v. Umbro Int'l., Inc., (Va. 2000).

PENAL CODE of Denmark, § 263.

PENAL CODE of Finland, chapt. 38, § 8.

PENAL CODE of France, chapt. III, art. 323-1.

PENAL CODE of Germany, §§ 202, 303A, 303B.

PENAL CODE of Hungary, § 300C.

PENAL CODE of Iceland, § 228.

PENAL CODE of Italy, art. 615.

PENAL CODE of Norway, §§ 145, 151b.

PENAL CODE of Poland, art. 267.

PENAL CODE of Switzerland, art. 143b.

Playboy Enterprises, Inc. v. Calvin Designer Label, 44 USPQ 2d 1156 (N.D. Cal. 1997).

Playboy Enterprises, Inc. v. Netscape Communications Corp., 55 F. Supp. 2d 1070 (C.D. Cal. 1999).

Price, D. A., "Exchange Trustbusters," *Business 2.0*, Aug. 22, 2000, p. 60.

Privacy Act, Austria, 2000.

Procter & Gamble v. Shanghai Chenxuan Zhineng Sci. & Tech. Dev. Co., Shanghai No. 2 Intermediate People's Ct. (2000).

Promotion & Protection of the Information Infrastructure Act, South Korea, 1999.

Putnam Pit v. City of Cookeville, 221 F. 3d 836 (6th Cir. 2000).

Radcliff, D., "Got Cyber Insurance?" *Computerworld*, Aug. 21, 2000, p. 44.

RealNetworks, Inc. v. Streambox, Inc., 2000 U.S. Dist. LEXIS 1889 (W.D. Wash. 2000).

Regulations of the People's Republic of China on Protecting the Safety of Computer Information, Chap. 4, Art. 23.

RIAA v. MP3.com, Inc., 2000 U.S. Dist. LEXIS 5761 (S.D.N.Y. 2000).

RIAA v. Napster, 2000 U.S. Dist. LEXIS 6243 (N.D. Cal. 2000).

Robinson, E., "Battle to the Bitter End," *Business 2.0*, July 25, 2000, p. 134.

Rohde, L., "U.K. E-Mail Law Reaches U.S.," *InfoWorld*, Sept. 4, 2000, p. 28.

Rongshu.com v. China Publ. House, Beijing First Interm. Ct., 2000.

Sanger, I., "Cyber Crime," *Business Week*, Feb. 21, 2000, p. 37.

Simons, J., "X-tracurricular Activities," *Business 2.0*, Jan. 23, 2001, p. 30.

Thibodeau, P., "Feds Set Financial Security 'Guidelines,'" *Computerworld*, Jan. 22, 2001, p. 7.

Thurman, M., "I Hired a Hacker: A Security Manager's Confession," *Computerworld*, Feb. 26, 2001, p. 50.

Ticketmaster Corp. v. Microsoft Corp., No. CV 97-3055 (C.D. Cal. 1997).

Ticketmaster Corp. v. Tickets.com, 2000 U.S. Dist. 4553 (C.D. Cal. 2000).

Unauthorized Computer Access Law of Japan, Law No. 128, Art. 3.

Uniform Computer Information Transactions Act (UCITA), at http://www.law.upenn.edu/library/ulc/ucita/cita10st.htm.

Uniform Electronic Transactions Act (UETA), at http://www.bmck.com/ecommerce/uetacomp.htm.

United States v. Microsoft Corp., 97 F. Supp. 2d 59 (D.D.C. 2000).

Universal Studios, Inc. v. Reimerdes, 2000 U.S. Dist. LEXIS 11696 (S.D.N.Y. 2000).

Urofsky v. Allen, 955 F. Supp. 634 (E.D. Va. 1998).

"Vive la Liberte," *The Economist*, Nov. 25, 2000, p. 75.

Wang Mong v. Century Internet Comm. Tech. Co., Beijing First Interm. Ct. (1999).

Washington Post Co. v. TotalNews, Inc., No. 97-1190 (PKL) (S.D.N.Y. 1997).

Wilder, C., and J. Soat, "A Question of Ethics," *InformationWeek*, Feb. 19, 2001, p. 39.

Yesmail.com v. Mail Abuse Prevention System LLC, (N.D. Ill. 2000).

Zeran v. America Online, Inc., 129 F. 3d 327 (4th Cir. 1997).

About the Author

Jeffrey H. Matsuura is an attorney with the Alliance Law Group in Tysons Corner, Virginia, where he specializes in providing legal and business counsel to enterprises involved in the information technology, telecommunications, and digital media industries. A member of the District of Columbia Bar and the Virginia Bar, Mr. Matsuura is the author of *A Manager's Guide to the Law and Economics of Data Networks* and coauthor of *Law of the Internet*. Mr. Matsuura has worked as counsel for several technology-based companies, including Satellite Business Systems, MCI Communications Corporation, Discovery Communications, Inc., COMSAT Corporation, and TELE-TV. He has also served as an advisor to the Virginia legislature's Joint Commission on Technology and Science and to the National Task Force on Knowledge and Intellectual Property Management. He has taught as a member of the adjunct faculty at the University of Maryland and Northern Virginia Community College. Mr. Matsuura has earned degrees from Duke University in North Carolina, the University of Virginia School of Law, and the Wharton School at the University of Pennsylvania. He can be reached at jmatsuura@alliancelawgroup.com.

Index

Visual Telephony, Edward A. Daly and Kathleen J. Hansell

Wide-Area Data Network Performance Engineering,
   Robert G. Cole and Ravi Ramaswamy

Winning Telco Customers Using Marketing Databases,
   Rob Mattison

World-Class Telecommunications Service Development,
   Ellen P. Ward

For further information on these and other Artech House titles, including previously considered out-of-print books now available through our In-Print-Forever® (IPF®) program, contact:

Artech House   Artech House
685 Canton Street   46 Gillingham Street
Norwood, MA 02062   London SW1V 1AH UK
Phone: 781-769-9750   Phone: +44 (0)20 7596-8750
Fax: 781-769-6334   Fax: +44 (0)20 7630-0166
e-mail: artech@artechhouse.com   e-mail: artech-uk@artechhouse.com

Find us on the World Wide Web at:
www.artechhouse.com